For Lisa,
With Love

THE GIFT OF INTERSPECIES COMMUNICATION

STORIES AND EXERCISES FOR THE SOUL

Nancy Orlen Weber

By Nancy Orlen Weber

Unlimited Mind Publications

Unlimited Mind Publications
THE GIFT OF INTERSPECIES COMMUNICATION
STORIES AND EXERCISES FOR THE SOUL

Library of Congress Catalog.

First Edition

Published by the Unlimited Mind Publications
(A subsidiary of NOW, Inc.)
P.O. Box 1132
Denville, NJ 07834

ISBN-0-96-46-118-3-x

Other Books and Products
by Nancy Orlen Weber

PSYCHIC DETECTIVE - True Stories & Exercises For the Soul

INNER JOURNEYS GUIDEBOOK

INNER JOURNEYS FOR MOTHERS-TO-BE
(Manual, CD & tapes for Creative Visualizations)

WORLD WITHIN & AWAKENING
(CD & Tapes)

Dancing Light Series
Blank Note Cards & Matching Envelopes of
Painted Illustrations

All products can be obtained through
Unlimited Mind Publications
PO Box 1132
Denville, NJ 07834

Introduction

I am not an expert on horses, cats, dogs, flies, snakes, trees, flowers, humans, stones, mountains, water, air, fire and so on. I still don't know the anatomically correct titles of the physical parts of horses except for head, mane, and hooves. What I am good at is throwing myself in whole heartedly – believing that if I clap my hands Tinker Bell will live. If I love deeply enough, and use myself as a transmitter of that love, the animal who is hurting will benefit.

From a pitifully shy child who was invisible for years to a joyful emergence took years of commitment to ripping apart my fears and discovering what was inside them. Like many other people, I discovered that other life forms won't laugh at me, will easily bond when given the opportunity, and once they love you, its for life.

Throughout my working life as an animal communicator, I have met people who know they talk with animals. Some believe that their cat, or dog, or horse only understands them. But everyone communicates with everyone at some level. Some animals, like people, are easier to commune with than others. Some need special assistance from people trained in their dialect, and others may not care to be aware of what's around them. This book is dedicated to all those wonderful people who have sheepishly admitted that they have conversations with their pets, all the caring volunteers whose love reaches the homeless stray who becomes their best friend, and all of us who strive to learn to love the world and be not afraid.

I hope that in the chapters that follow you will find ideas, exercises and models for even greater communion between you and all those other sentient beings you love. As one of a number of animal communicators, I feel fortunate to be writing at a time when there is growing interest and acceptance of communication between species. Today, we have such talents as Pamela Hannay, recently deceased, author of Shiatsu For Horses and a truly gifted teacher, Penelope Smith, who so many of my friends and students have gone to and love, and Nancy Regalmuto, another truly gifted psychic and healer, who years ago attended one of the earliest Horse Clinics Pamela and I performed. Nancy and I were both featured in Equs; being older, I was first.

What I love most is to see people's light shine brighter. It happens in a brief interlude of the shedding of an old belief and the acceptance of an expanded one.

Some of the beautiful transformational moments for myself have occurred during heart-talk. That's what I call what happens between myself and an ant, a fly, a grasshopper, Boo Girl, Sweetie Pi (correct spelling – she was actually Sweetie Pi R squared), Fiama, Barney (he was not purple), and other sentient beings.

Because I believe that the way we all learn is to do, along with the stories are suggestions on becoming an animal communicator.

This is my wish for you:
To see all of life as the extraordinary art of a master with a plan,
To develop deeply loving relationships with other humans and all other creatures of this world,
To feel the faith, love and trust that happens when we experience the invisible world of heart-talk.

Feel free to contact me at my email address with questions: LIGHTWING@aol.com or write to me at: Box 1132, Denville, NJ 07834. I would love to hear your precious truths in this, the Aquarian Age of brotherhood.

With Love filled Light,
Nancy

Dedication

This book is dedicated to Sweetie PiRsquared, born of a Bronx Jewish Deli grey striped feline who fell in love with a Puerto Rican beauty in the Spanish grocery on the Grand Concourse, Bronx, NY. She was one hour old when we met, laying in her mothers front paws, cuddled like a beloved child should be. Sweetie Pi slept on my pillow for 18 dreamy years. When my husband Dick entered my life, he, too, embraced this feisty, spiritual feline who accepted both of us as her companions – providing she had first dibs on the bed. She now escorts other departed souls into the nonmaterial planes. After Sweetie Pi's departure from earth, Boo Girl, a feline feral rescue, finally slept on the bed and at twelve years of age broke through her own fears and began introducing herself to all our friends. Boo Girl will explain everything later in the book.

To my children who spent hours helping rescue animals feel safe; Harriet, dear friend with a heart as big as Texas, who believed in my ability to help her rescued horses; and Pamela Hannay who collaborated with me on horse clinics, Sonya Oppenheimer whose affinity for rescues is delightful to behold, thank you for being my teachers, students, friends, and trusting souls. Pamela, though no longer with us on this earth, continues her teachings through her book, "Shiatsu For Horses," and the curriculum she helped create and teach as senior instructor of the Ohashi Institute for Shiatsu in New York City.

To Phyllis my soul sister, who helped me believe in my gifts and took me to the paddocks, to Linda Hepburn who introduced me to Amir, the first horse I worked on, you are both now in that mysterious universe beyond death. I pray that your souls are at peace and your families, who dearly miss you, lovingly feel your presence whenever needed.

To all my students who brought me their problems with animals and to all those animals over these last twenty five odd years who have allowed me to work with them, find some of them when missing, help with their need of healing, and grieved with their friends when they have gone on to the great adventure on the other side, I thank you all for trusting me to help. May I always live up to that trust.

Hannelore Hahn, founder of the International Women's Writing Guild, thank you for being a guiding light to so many of us.

Boo Girl

All our fears
of all others
are a reflection
of our fear
of the
power within.

Harness the
power within
and we are at peace
with ourselves
and
with the
world we live in.

Contents

Illustrations - Nancy's Doodles

Can we ever feel what a cat feels?

Tina and Jesse
reading their favorite story

Hatching

Beginning to Heart-Talk

Little girls from Brooklyn have cracks in the pavement to watch. Ants have little girls watching them live in the cracks, build their homes, help their friends and – if they like her – a trip up her arm. That's what I recall when I relate my earliest experiences with other species.

I could sit and watch this miniature world and have them lightly tickle me for hours. Being around the ants taught me how one species could trust another. They were not afraid of me and in fact used my arm as a bridge from one place to another. Their trust that I would not harm them is continuously honored. I walk as carefully as I can, stepping around the smaller creatures. There is always that little girl inside me: talking, remembering, and thinking how big I must appear, how loud my steps must sound. Most days, before stepping outside, I call out in my mind to all the creatures to warn them that this bigger than them being is about to enter their world. That to me, is Heart-Talk.

Mom and my sister, Anita, were terrified of things that moved unexpectedly – insects, snakes, worms and other wiggly creatures. One day, when I was seven and my sister was twelve, we were taking our clothes to the laundromat a few blocks from our home. We met a boy who showed us four kittens in a carton; he asked us to please take one. We did. We went straight home eager to play and love this little ball of fluff. Within hours, the kitten had found a warm cozy place somewhere in our home. That somewhere was a mystery to us. Mom became terrified that something happened to this tiny kitten. Eventually, like every kitten I've met, this one awoke and immediately began to chase everything that moved. Mom couldn't handle the responsibility and ordered us to return the kitten immediately. Fortunately, the boy was still there. In desperation, my sister and I begged for some other living being to play with. Mom and Dad agreed to purchase a bird. I don't know who named him Moses, but that's what we called him. A few months later, he met his maker on Yom Kippur; cause of death unknown. Eventually, my sister tricked our folks into letting us have a puppy. That was Suzie, my buddy. She listened to my fears, anger, joy and sorrow. When I cried, she licked my tears, when I giggled she jumped up and down in glee. Suzie talked the language of love and *that* I understood.

As an adult, I have more than made up for my childhood lack of furry and feathered friends. Ask Rebecca and Jesse (daughter and son, respectively) about the twenty two kittens I agreed to temporarily house in our home. It would have been fine if I knew to give them (the kittens) a flea bath as soon as they entered our home. Rebecca and Jesse grew to love and keep several of these kittens; yet, their clearest memory is of scratching flea bites all over their own little bodies. I'm still apologizing.

Enhancing the gift of communication:

Treat every living being as if there is a divine presence operating the life force within the form. Approach each sentient being with a consciously open heart and a flowing mind. Imagine your personal mind is one track with an overlapping universal mind track that is not just connected to all else, but is all else. Acknowledge; asking for the universal mind to vibrate at its highest divine resonance as preparation for your conscious connection with other life forms. Allow your higher self to observe the depth of the interaction between you and another.

It was 1976 and I was invited to Andrija Puharich's Monday night class. Andrija Puharich, now deceased, was a leader in the field of parapsychology. Dr. Puharich and Aldous Huxley used a Faraday Cage for their research in telepathy. Their findings are in "Beyond Telepathy" by Dr. Andrija Puharich. Most people are more familiar with the psychics Andrija discovered and brought to America – Ingo Swann, Peter Hurkos and Uri Geller.

This Monday Marcel Vogel was a guest of Andrija's. Marcel Vogel, also deceased, is considered the father of the study of crystals and liquid crystallography in our country. He was a pioneer researcher at IBM and had the first television show on metaphysics.

Marcel called me a few hours before class to introduce himself and to play.

"Nancy, can you tell me what I am holding in my hand?"

"Marcel I see a glass covered small object. It's round and it ticks."

"Nancy that's wonderful, yes I am holding a watch."

When we gathered in Andrija's living room for class Marcel was invited to teach. Marcel took a potted plant and placed it in the center of the room. He then guided us into a meditative state. We were instructed to place our hands near the plant and imagine our energy traveling through our fingers into the roots of the plant. As we continued this exercise, exploring the stem and the flower, it was easy to set aside the part of the mind that says this is silly. After the meditation we all shared our experience. It was interesting to observe that most of us felt the plant distinctly unhappy with its windowsill home and wanted to be moved to another part of the room. Andrija happily complied and took the plant to where we believed it would be comfortable. Several weeks later it was apparent that we heard the plant. It had never produced blossoms before, it became beautifully vibrant with colorful flowers.

Let go of the need to figure out what you are experiencing until after your experience. The left hemisphere of the brain is always seeking to rationally understand. It's your turn to play.

Journal of observations:

My earliest memories of an interest in interspecies communication is

Revelation:

After practicing my own centering techniques and sending a universal message of oneness, the difference about this encounter is ...

To have
a wild animal
love you
Is to know what
honor, respect, and love
is about.

**Our beloved furry friends
are constantly communicating
with us,
wondering when
we are
going to answer.**

Heart-Talk With Baby and Boo Girl

Communicating with other species begins with a spiritual intention. I think that if I were some other kind of animal, my intuition would be keener. I wouldn't be arguing with anyone about whether my senses were accurate or not; I would be depending on them. We humans wrap ourselves in a fabric of belief/disbelief, acceptance/argument, and intellectual debates about existence while Baby, a most precious groundhog, lives in the moment. We call her Brahma Baby.

My love of ground hogs developed through Boo Girl's friendship with The Trio, as I knew them. In 1981, Boo Girl's mommy had abandoned her seven children on the grounds of a local institution after a fumigator commenced war on their home. Fortunately, the fumigator saw the babies and called the local animal warden. Seven stranded kittens waited for help.

My girlfriend, Sue, had a shelter where I had already obtained our dog, Ramona. That summer, Sue needed a break and had taken me up on my offer to help, "Anything you need, just call." If you say those words, remember there is an occasional person who will believe you. Sue asked me to take in all the rescue kittens for the summer.

Seven wild kittens were brought to our home in a cage, hissing with the tiniest set of claws and teeth. We put their cage in the red room. It is a 25'x25' living room that adjoins another 25'x25' living room through french doors. All this second living room contains is white walls and a red carpet. It's the room in which the children loved practicing gymnastics, bringing wizards spells forward with a slashing wooden wand and dancing. Although we have since moved and others have now probably done away with the beautiful emptiness of the space, I still see it as the room that always waited. So the cage found a home, at least until all seven babies felt safe enough to come out and play.

The first one out of the cage spoke lamb, the sweetest, gentle baa-baa sound broke through that fur ball. She's still a lamb speaker; no meow yet and she's turning 19 years of age. My children decided to name her Boo Girl and claim her as our own. The other six were taught to trust humans through the patience and compassion of Rebecca and Jesse. They would lie perfectly still for hours at a time in the same room with the terrified kittens. After almost three weeks, all but one would happily climb on the mountains and play. As they became accustomed to playing on the backs of two still children, hands would gently reach out to stroke them. By the end of the third week, all were relating to humans well and Rebecca and Jesse could move around enjoying the abundant delight that was now present in our little friends. We unanimously agreed that while Boo Girl was the friendliest she was also the most terrified. She would cuddle and then bolt at the slightest movement. If I called out, "Boo Girl, it's only mommy," she would pitifully baa-baa her way back to one of our waiting arms. One baby, Maxine, was able to escape into the woods. For three months she stayed there. We left food outside, bringing it closer and closer to home, hoping to lure her back. Jesse, age 6 at the time, was the hero. He would stand outside calling her to him. One bright summer day, it happened. She placed her long calico fur under the palms of his hands and purred. She was back with us. Now they were ready for new homes. Boo Girl's siblings were easily given loving homes through the shelters adoption screening program. Wise decision, since we had five other grown cats who were quite content with their established commune.

Boo Girl took to our tiniest family member, a silver tipped tabby, Tina. She was our yogi master. Until Tina's death, she could do yoga postures, including lying on her back in your arms with complete devotion to whomever she was with. Beauty, a rescued morning dove, loved flying onto Tina's back and perching for a picture. Boo Girl, like everyone else, loved Tina. Tina loved everyone and would happily bring Boo Girl on her side of the bed, far away from the evil eye of Amy. Amy, our largest feline resident, believed that she set the rules. And within her claw reach, she did. She loved people of all ages, especially children. They could scrunch her up and she would purr louder. Her claws were reserved for felines, all except Tina. To all who met her, Tina shared her heart and soul.

Boo Girl quickly decided that both indoors and outdoors had opportunities for pleasure. She began hanging out by a boulder. Okay, it was a large rock, but I couldn't budge it. If you opened our front door and walked ten feet across and about six down the hill towards the street, you would come to this oversized rock that would need blasting to be moved. This was a most intriguing place for her. One day, I looked out and saw Boo Girl sleeping with three ground hogs on the rock/boulder. It was impossible to disguise her; she looked like a miniature bobcat, warm browns, splashes of burnt orange and black, with fur soft as a lamb. It went well with her sounds. Neighbors teased me when recounting this spectacular sight. Several weeks later, two neighbors joined me in a hunt for great clothes at end of the season sales. To return and enter my driveway we had to pass the front yard. There, four ground hogs were dancing playfully. The car sounds disturbed their musical ears and they disbanded, all running in separate directions, three towards the woods and one silly one towards my door. You should have heard the talk in my neighborhood.

Boo Girl continued to enjoy many good years of friendship with lots of ground hog families until we moved to another town. In our new home, we lived where the land developers had destroyed the woods and left food for grazing, or lawns, in its place. Why do we call it land developing? Isn't it land destroying? Now, I want a home and I want it built on land, but just like our euphuism, "we put her down" or "put to sleep", no, I don't think so. We chose to put to death or kill. Perhaps it is humane in some circumstances, but lets call it what it is. A builder destroys the land, at least a piece of it, sometimes killing small animals, certainly trees, plants, bushes and the homes of many insects, birds and animals. It may be necessary, but I don't think it has to be as destructive as it gets. And what about the native (not just North American native) way of thanking the land, telling the trees and the other living beings what we are about to do, giving thanks for their presence and acknowledging a choice we are making to change the contours of this established earth, giving all the living connected energies a needed moment to adjust to the changes?

Because we don't openly recognize the flow of energy that connects all of life, we act accordingly. We cut off and are without depth in our deeds. Of course, not in every way, but when was the last time you went to get a Christmas tree or called a tree service to take down some trees and before doing so gave thanks to the life force, gratitude for their presence and explained to them what was happening? If it sounds odd, it is because it has been drummed out of us. Are we supposed to pray and not mean it, to donate to environmental causes and not be consistent with our beliefs? Anytime I moved to an area that could prove dangerous or upsetting to one of our animal friends, I would sit down and send them images along with emotionally charged thoughts of the forthcoming changes. This is as important as telling my children we are moving. Every being needs to adjust to changes that are beyond their control. Boo Girl seemed to understand and took to walking around the cul de sac late at night, shy thing that she was.

Around that time Janice Hall, a friend and astrologer wanted a reading. After her reading, she brought out my natal chart and offered to reciprocate.

"You will be with your soul mate in two years. He's a Capricorn, music will have something to do with your meeting, and he will bring organization to your world like you have never seen. You will love him completely and forever."

I wasn't pleased. Disinterested was the only word I could think of to describe a permanent relationship. I was now happily strong, independent and very content being single. I left hoping she was wrong.

But she was right. And Boo Girl benefitted from the relationship I had told myself I didn't want. Furthermore, it all came about because of music. I had collaborated with a friend, Elaine Silver, in writing twelve songs. Elaine is a gifted folk singer-composer, about whom Tom Chapin had glibly remarked, "I'm her closing act." Two of our songs became some of Elaine's performance favorites.

I was always humbled and delighted to hear Elaine sing my words. Mesmerized by the experience, I became something of a groupie. When she said that she was opening for Loudon Wainwright at the Stanhope House in Stanhope, NJ, I took advantage of another opportunity to hear "my" songs.

Another friend of Elaine's, Dick Weber, was also in the audience. He introduced himself and I immediately took to disliking him. I continued to run away from him as we re-met at various environmental functions until two years later when Dick joined a nonprofit that I had started with another girlfriend, Sue Velicoff Pelechaty. Thinking and swearing that I would neither fall in love again nor marry again, I landed in Dick's arms and have stayed there ever since. See how psychic I am?

We – Boo Girl, Tina, Amy, Sweetie Pi, my son Jesse and I – all moved in with Dick in his ranch style home in a lake community. With beautiful woods behind our home, it is the perfect place for wildlife to find peace. Boo Girl promptly took to sitting for hours outside on a stone wall waiting for Baby.

Baby, a beautiful groundhog had been born that year and mom left ASAP. Baby was as disinterested in a relationship with Boo Girl as I had been in one with Dick. She would not even look at Boo Girl. So, I started to leave food out for her; you know the Jewish mommy bit. Baby liked melon. That was a major breakthrough. A piece for me, a piece for Baby. Boo Girl would drool, not for melon, not for food, but for friendship.

After weeks of Baby eating on top of a three foot wall behind our house I knew the moment had come. Everything was different; the air, the feel of the earth. My thoughts had stopped and I followed my heart right out the back door, across the yard and stopped beside Baby. My right hand reached out and stroked her beautiful back while she continued her melon binge. Her head turned, eyes directly on me, a moment of acknowledgment and then she simply continued to eat. My hand brought itself back, prompted by some distant divine knowledge that my conscious mind wasn't privy to, and I stepped back slowly until I re-entered my home. Baby finished her dinner and turned back up the hill to her sleeping quarters. A few friends had witnessed this precious connection, Boo Girl being among them. Immediately after that encounter Boo Girl introduced herself to each and every student in my at home Thursday night class.

"When did you get this precious cat?" Linda asked as she petted this new friend.

"Twelve years ago."

Boo Girl proceeded to introduce herself to each and every visitor, cuddling sweetly in their arms.

Several months after completing this chapter, and six years after she came out of her shell, Boo Girl died peacefully and sweetly in my arms.

Exploring the gift of communication:

Enter into the situation without preconceived notions and with a willingness to be available to the highest spiritual outcome. Like a good research scientist, you may have a hypothesis. The illusion is limiting, the truth has no limits. We do not have to know the “how”; we need to simply enter into the situation without preconceived notions and with a willingness to be available to the highest spiritual outcome.

Choose some other species to become friends with. Know in your heart that this is possible, even if it does not appear probable. Every day, whether you see this friend or not, place a greeting in your mind/heart and send it to this friend. Image your friend receiving your blanket of love. Trust your intuition if you are moved to do some unusual actions in regard to this new friend. Remember simple truths; if it is harmless to you and others, do it. There is a man in our county renown throughout the country as one of the leading experts at Raptor surgery. He never went to vet school; he never took training. He simply opened his heart to a Raptor in trouble years ago and did what his spirit compelled him to. He did not argue with his inner voice; he did not argue with his circumstances, he did not consciously seek to become an expert with Raptors. His soul took him there. He complied.

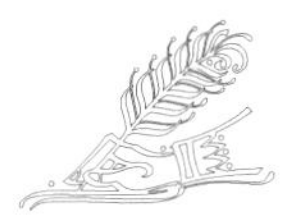

Record your amazing experiences.

Journal of observations:

One of the most unusual experiences with other species was on (date)___________ and occurred (where)__________________.

This is what I remember

Exercising Heart-Talk:

Mentally talk to whatever living beings exist in the same environment as you. Plants, insects, mammals, etc. all are part of the vast sea of intelligence. Send a loving thought along with a mental image of the thought to the plant, insect, etc. Imagine your heart sending the love and happiness at meeting another creation of life. Get into the feel of the awesome mystery of all of us being somehow connected. Be grateful for the opportunity to connect without assuming you must receive a clear and precise answer. Let go. Enjoy. Wait. Accept whatever connection is established as a binding contract between two spirits, each acknowledging the other, as a manifestation of the spirit of all life. Keep practicing. Eventually it becomes what you believe and belief creates existence.

My experiences are...

There are great souls
in every species;
Namaste is true for all;
My spirit greets your spirit

Michael Returns

Spiritual intention is everything; Mu, an exquisite, long haired, male feline was one of my gurus. I don't know if he was an atheist, pagan, or God centered being, but all his actions demonstrated that he lived within a code of ethics filled with integrity, humanity, compassion, kindness, humor and love.

Back in 1974, Sweetie Pi gave birth to a beautiful long-haired black guy we named Michael. He was killed at eleven months of age and his burial was attended by his family – his mother and her best friends. In the ancient tradition, we took artifacts that we believed were special. Rebecca gave him a favorite toy to take on his new journey; Jesse gave him one of his toys; I gave him flowers, and Sweetie Pi cried.

Michael had been a difficult child for Sweetie Pi. Born after a C Section, she, like her mother, held him in her arms and cried. She shed more tears in her life then some men I've known. When it came time to wean him, she would not hear of it and stalked away yelling at all of us. She would lick off whatever substance we placed on her nipples so that her son could continue nursing, until she decided when he was six months old, that it was time. The Veterinarian kept insisting she would develop problems, but the only problem she developed was Michael's lack of hunting instincts. One day, she grabbed a mouse and put it in the water bowl. She stalked across the slate kitchen floor and talked to Michael. We sat at our cane woven kitchen chairs watching her direct him to the mouse. Michael went over and took the mouse out of the bowl, gave her/him a few licks and proudly smiled at mama with a "aren't you proud of me, I saved a mouse's life." To Michael's surprise, mama would not play with him the rest of the day.

A few days later, Sweetie Pi tried again, this time, out on our hilly lawn. The grass was tall enough for her to pretend her white with black markings coat could be camouflaged. No one argued with her, so we let her believe it was a great technique. It must have been, she could catch anything – except when Michael stalked her. He was not quiet. He tromped after her, jumping with ecstacy at the new game. He destroyed any possibility of follow through with the loudest meows. Sweetie Pi threw up her paws in frustration and never tried to teach him again. However, love is powerful and when he died, she, like all of us, mourned his loss.

A little over one year slid by and there was a storm rattling our windows. I went to look outside and there was an all white long-haired feline rolling around at our doorstep. This became a daily occurrence. He would not enter our home for the first few weeks, but he would eat as if he were starved. The odd part is we discovered he lived a few streets down, seemed happy there and just took off to our home and decided to let us adopt him. The family who lived with him concurred, it seemed the only thing to do.

I didn't think anything of this, being a little slow sometimes, but I was amazed when Sweetie Pi had absolutely no animosity. This was the Queen and she had said, "Yes, Yes, Yes." From the moment they met, they were inseparable, except when she went to hunt. He followed her everywhere at first; then, they reached a compromise. He would stay inside when she hunted.

The reason was obvious. Birds loved him; mice loved him; dogs loved him; he had no enemies and he loved them all. We named him Mu, for Lemuria in honor of the wisdom of the ages that poured through him as naturally as water down a fall.

Mu liked to lie outside in the early morning and be among the Morning Doves as they ate on our porch. They always accepted his presence and just walked around him. Once in a while, another species of bird would dance up above his head and he would just sit there being entertained by their beauty.

One sunny spring day, a neighbor called asking if her mom could stop by. "My mother is a member of the Audabon Society. Every time she passes your home she sees birds that she hasn't seen in Westchester County in years."

So mom dropped in. Flowery dress, sweet voice and gray hair accompanied this happily anxious woman as she opened my screen door and sat facing the outside like a true bird watcher. We spoke while she kept her eyes on the air right above our porch.

"Look, a Titmouse." I kept looking for a mouse with you know what. So much for my species identification abilities.

"That's the very pretty grey bird to the right." She was used to explaining to children like me.

Mu stood in front of me; then, walked to the screen door. I got his message and opened it for him. A few seconds passed and he was ready to come in again. In his mouth was the Titmouse, gently held with it's eyes open and peaceful. Mu went and stood in front of the woman who promptly screamed.

"Mu, she doesn't understand. You'll have to let him go back outside." I opened the screen door again and Mu walked, head high to the porch.

"Please stop screaming; he just wanted to help."

She stopped all right. Mouth opened wide, she watched the Titmouse fly within inches of Mu's head for a few minutes while Mu took his "I am a sleepy cat" position, fully stretched. I could swear he and the Titmouse were laughing at the absurdity of someone being frightened for one of two best friends.

Again my neighbors talked.

Mu enjoyed a good rescue. Being the responsible, evolved soul, he felt it was his duty to help wherever he could. I had come to rely on Mu's common sense when we adopted Creamsicle, named for his coloring. Creamsicle was a male feline, abandoned and brought to a shelter. From the beginning, he seemed a bit strange. Although he was less than three months old when we adopted him, he began to wander off immediately. Mu would know where he was and when I called out for Creamsicle, Mu would search and rescue.

Marianne was our babysitter. At 15, the eldest of five children, she carried responsibilities well. She would also, when the need arose, feed our feline friends. So, when we were going on an overnight trip, Marianne stopped by for instructions.

"Marianne, if Creamsicle gets outside and you can't find him, go to Mu and tell him to get Creamsicle."

"Uh huh"

"Marianne, you don't have to believe me, just do it."

"I promise."

The next day all seemed quiet, Creamsicle and Mu asleep together on the living room sofa. Marianne stopped by in the evening after her basketball practice.

"My family can't stop talking about Mu. Yesterday as I opened the door to get in, Creamsicle dashed out and up a tree. I turned to Mu and said, "Mu, Creamsicle ran up a tree, please get him. I didn't raise my voice. It was just like a conversation with a friend. Mu ran out of the house and dashed over to the tree where Creamsicle was residing. Mu climbed the tree and I swear he was yelling at Creamsicle. When he got to Creamsicle, he jumped over him and bumped him downward. He kept doing this until he got him down to the ground and then continued to yell at him as he poked and pushed him into the house! How could he know?"

And so Marianne opened her mind's heart to more wondrous possibilities. Hopefully, that experience has enriched her life. It certainly has mine. Mu has brought so many treasured moments to my heart. My prayerful wish is that his soul is continuing its grand adventure somewhere, somehow, in the divine light of creation.

The best homework is creating the belief and learning to live within it. Some of the fabulous books that have inspired and encouraged so many of us to do our work are: ***Kinship With All Life*** *by J. Allen Boone (yes, a relative of Daniel Boone), and* ***The Parrot's Lament*** *by Eugene Linden.*

When we moved to New Jersey in 1979, we lived in a large stone house up on a hill. Down the hill was a busy road, Route 46. Sitting with Mu and Sweetie Pi, I mentally drew picture after picture of the road, cars and squished cats. Along with the images, I created a feeling of terror, grief and pain and threw that into the psychic pot. Shooting this along the psychic highway from my heart and mind, Mu and Sweetie Pi acknowledged the Heart-Talk by sending me back oceans of adoring, affectionate, mommy thoughts. My whole being felt satisfied with a deep inner sense of well-being. They understood never to go to the highway.

That same afternoon an elderly couple called us to come get our cat. Their house was at the corner of Route 46 and our street. Alongside their cottage was several acres of woods that led all the way up to our lawn. Mu had been exploring. Rebecca was selected to bring Mu back home.

The next day another call from the couple and again Rebecca retrieved Mu. The third day I figured it was time for a meeting. We met outside their door.

"We have 30 cats and we never let them out. They could get killed."

"Mu is not suicidal. He's just checking out his new surroundings."

"You really shouldn't let him out. Now we'll never get him back, he went into the woods. We couldn't hold on to him."

"Don't worry, if it happens again, as I'm sure it will, just talk to him. He listens. Look, I'll show you. Mu, mommy's here; please come out sweetheart."

Mu scampered out of the woods, rubbed my legs and looked up at me.

"Mu, my love, these people are afraid you'll get hurt; they don't understand. Please go home and I'll try to help them understand. Ok?"

The poor couple just kept watching as Mu trotted back up the hill, obediently seeking our door where the children were waiting.

"See? All you need to do next time is call him and tell him to go home. If he doesn't listen to you, just call me and I'll send him a thought."

The next day our local newspaper had a photo on the cover of Mu, Sweetie Pi, our dog Ramona, Rebecca, Jesse and me. It was about the work I had done recovering missing animals. The headline was *Psychic Communicates With Animals.* I always wondered what the couple must have thought when they picked up the paper. They never called again. *Guru Mu* taught me to stretch beyond my beliefs and seek common ground with all animals.

Record your encounter with an animal who taught you to expand your beliefs.

Can we learn the language of a dog, a cat, a bird or a dolphin as easily as they learn ours?

Nancy, Beauty and Tina

The Other Side

For the past nine years my third week of August is spent in Saratoga Springs, New York, on the campus of Skidmore College. It is the site for the annual conference of the International Women's Writing Guild and the spiritual time for magic. Magic happens each year differently.

Sandy Van Hoose (now Sandy Van Hoose Saunders) came to my workshop at the Guild and signed up for a psychic reading while at Skidmore. Vibrant, outgoing and creative she quickly struck up a friendship with me. This is not unusual at the Guild; by the end of the week we all go home with friendships deepened, new bonds created, and inspiration to write, write, write. Sandy called me in November.

"Remember you told me that in November I would get news from the west coast, and that I would do something out there around Christmas that would change my life? I sent a letter to the NBC show *THE OTHER SIDE* about my standard poodle named Murphe being the reincarnation of my previous dog, a great dane named Murphy. They just called me. They're in Los Angeles and they want to fly me out around Christmas for the taping of a show on animals. I just wanted you to know. Do you ever work with animals?"

Now, that is funny since I always see the connection between my supposedly varied careers. Sandy and I had spoken about her dog but I hadn't thought it necessary to use my abilities in that direction since there was nothing wrong and Murphe wasn't missing.

Sandy had a wonderful experience about which she had written "LOVE RETURNS". Sandy sent the story to the producers of *THE OTHER SIDE.* Sandy's beloved Great Dane had always skipped one of the front steps (the same one every time) as he bounded up to the front door. After he passed away, Sandy searched for another Great Dane. When her brother's Standard Poodle gave birth, Sandy felt it was the right one – sight unseen, many states away – and claimed it as her own. Upon picking up her young, new friend, he promptly fell asleep on her lap on the way home. When arriving at her home, he stopped at the same step, jumped over it and headed to the favorite spot of her previous friend! He's been doing it ever since.

"I'm going to call the producer and ask if they would want some information on you. I'll call you back"

And call she did. And call they did. With my plane ticket in hand, I kissed my children and left for the west coast with the knowledge that I could still celebrate the holidays back home. Two days before Christmas, Sandy and I met in Los Angeles at the hotel the film producers had reserved. A delightful dinner was shared and then we went to her room where Murphee was staying. He is a beautiful black standard poodle. Sweet and friendly, his heart is very certain that Sandy and he are soul kin.

The next day, we were taken to the studio. Walking through the corridors we met Jay Leno on his way to work and Mel Torme finishing work. The show was put together by believers, from the audio engineers to the host. It was a wonderful experience and many of us who loved the show wish it were back.

Murphe, Sandy VanHoose Saunders & Nancy O. Weber

Exercising Heart-Talk:

Ever wonder who your cat or dog was? Ask them. Verbally and mentally send the question. Come on; it won't hurt and no one else needs to know. Of course it could lead to lots of interesting associations. And it can explain how some of our pets are strays that show up at our doors or how we can have the sudden urge to go seek an animal. Perhaps they are calling us to them.

Conversations with my animal friend

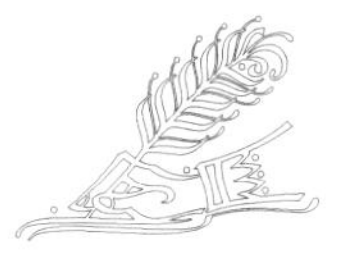

Exercising Heart-Talk:

Take a walk in nature. Consciously open your heart and mind to be receptive to all of nature. Send a message to the forest, the ocean, the sand, your backyard – wherever you are – "Creator of all life, my soul reaches for the light and through the light I reach out for all others. All of life is my family and I thank you for this opportunity to engage my expanded family in conversation." Believe the statement. Believe and be receptive. If you feel inclined to stop and watch an insect or listen to a bird, that is the member of our family who is responding to your message. Start talking.

While exploring nature I encountered 

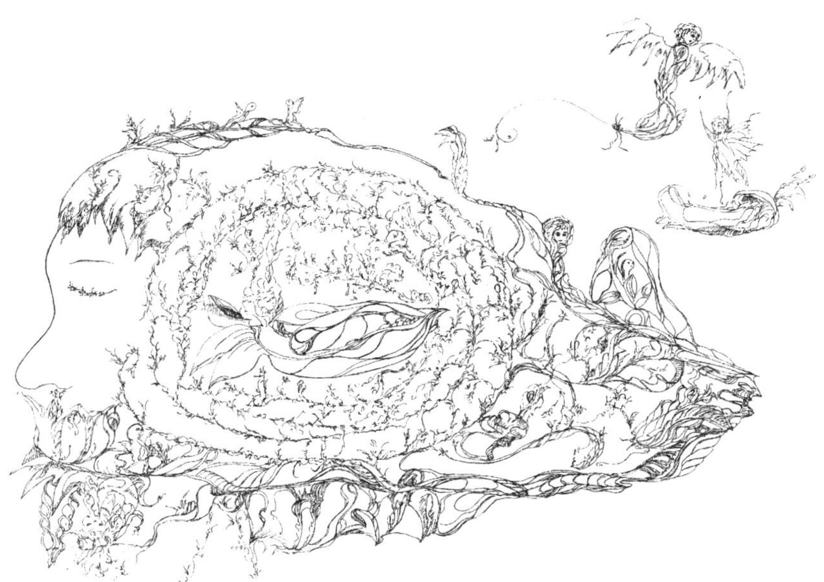

This life,
This dream,
This creation,
We are part of;
Just as the horse is part of;
and the tree,
and the bird,
and the stone,
and the grass,
All together form this life,
This dream,
This creation.

Be like the animal you love, mimic them until you can catch a glimpse of their reality.

Beauty Enters Our Life

It was a warm evening, 6:40 p.m. to be exact and the couple my husband Dick and I were talking with was contemplating their wedding.

This was the first time we were meeting with them. Dick and I each perform wedding ceremonies. When we can, we like to sit in on each other's interviews. Hearing couple's stories of their first meeting, when they fell in love, funny proposals and all the wrappings of relationship happiness delights both of us. We have met fabulous people from all walks of life this way. That night was a double treat.

Shari and Sean were delightful to talk with. They were saying something when I noticed out of the corner of my eye a flash of something dropping out of a pine tree on our lawn. Dick saw it at the same time and left the porch to walk over to the tree. I saw him bend down and scoop something. His mutterings were beyond comprehension but it sounded like he was upset.

Holding out his hand to me, I took the tiniest bird I had ever seen. It was somewhere between three and five days old and as we later discovered, a Morning Dove. Thank you Mu, our dear cat who loved birds, you sent us a Beauty.

Dick's only comment: "It's a very baby bird and it will probably die."

"Not if we can help it! Hurry, while I hold her, you go to the pet store, and get her some food."

Dick drove to the pet store, getting there just as it was closing, (a very good omen) and returned with baby bird mush, bird syringes and instructions. I tried to feed Beauty. Like most mothers I was awed by her appearance. Feeding her was another matter. Every fifteen minutes to half-hour she had to be fed, around the clock. Not Beauty; this mommy couldn't do it. I fell asleep with her in my hands and prayed that she would survive the night. Mentally covering her with prayer, light, love and anything else the universe could provide, I awakened to a starving child. Did you ever try and feed a Morning Dove? They don't open their mouth like other birds. In fact, they keep it shut and toss their head side to side to spit out the food. That was day five or six. That's when I knew she would survive. Just like some of us when we start recovering our strength - ornery.

After several days of battle with our stubborn infant, Dick and I were sure that this little bird did not fall from the nest, she was pushed.

As we later learned, doves are raised on “crop milk” which the parents force down chicks’ throats. So conventional wisdom tells the aware not to bother with hopeless causes. Our lack of awareness was another very fortuitous circumstance for our new “Beauty”.

Beauty slept in a basket, then on our dresser. She stayed in my hair during the day or on my shoulder. I tried to teach her to coo, but she still ignored my attempts. Dick taught her to fly, running with her on his arm and then gently flinging her in the air. She took to eating on our living room floor, kitchen table and porch. Tina, our silver tipped tabby, fell in love with Beauty. Beauty reciprocated by sitting on Tina’s back and letting Tina be the protector when out on the porch. Our other feline friends didn’t care either way. Boo Girl wasn’t interested in anyone except Baby, the groundhog, and Amy, our fourteen pounder feline, wasn’t getting up for anything less than a hug or food.

Our neighbor called out one day, “Is that your bird who visits us?”

“I don’t know, what does it look like?”

“It’s brownish grey and small with a young face. I opened our door yesterday and the thing just walked right in past our four cats, into the kitchen and up on the table where it proceeded to eat the leftover cereal.”

That’s our Beauty. Isn’t she wonderful.”

Mama was so proud of her. I still rejoice in the image.

A few weeks later, Beauty was flying, crooked as a Candyland path, but flying she was. She would stay out for a few hours, then strut back in past the open door, jump up onto my shoulder and busy herself watching the world. One night, while teaching meditation in my living room, all heads turned in unison. Beauty was staring in, perched on a rope that was lying above our outdoor wooden swing. I opened the door and she few in, landing on the telephone by one of the students. We took it to mean an important call would be coming for this student. Then she flew across the room onto my shoulder where she remained peacefully joined with us in meditation.

That was Beauty’s last night with us. She slept on our dresser, leaving a string of "pearls", and the next morning played outside watching mama in the garden. In the midst of my weeding Dick's myrtle, (he hates anything getting in the way of its grown), Beauty flew onto my head. I felt her bid me goodbye as she flew across the garden, across the road and into a very old, very high tree far beyond my reach. Another child was leaving home.

When my son went off to college, we drove him up to Amherst, parked his stuff, had a happy time with him and all the way home I played harmonica and sang, happily knowing he was going to try to fly a little at a time.

Beauty's leaving was different. I didn't feel as safe. My fear level was high and I couldn't sort out what it indicated. Learning to differentiate between fact and fiction, imagination and psychic perception, is a continuous process.

The next day a storm erupted, one of those tearing storms that pulls telephone wires across trees, breaks huge branches and rips roofs. We were fortunate and there was no apparent damage other than a few hours of peaceful silence while the electric company reconnected downed wires. Dick and I began a fearful "shall we mourn her loss" pilgrimage to all Beauty's known haunts. No Beauty. No sightings. No peace.

A few months later Dick skipped up the front brick walk.

"Guess who I saw – Beauty is alive."

"Where? Are you sure?"

"Who else would do this? I opened the basement door and went inside, a few feet in I turned to look at something and there she was following my footsteps. When I reached down, she took off, but she didn't make any sounds of cooing and she flew crooked! It had to be her. Who else flies like that?"

So Beauty still lives and we can easily spot her. She has a mate and he flies well. Occasionally she pays us a visit, walking in our garden and sitting on a feeder by my office window, staring at me with those beautiful dark eyes that say so much. It seems that all Morning Doves treat us with the same friendship as Beauty. Anywhere we are, they don't fly away, they walk right near us. If they are feeding by my window and I open it, they continue with a little friendly nod.

I can't say that I actually ever heard Beauty other than her mommy greeting and mine back.

"Beauty, mommy loves you" is said with a long cooing sound attached to it.

The spirit of all is connected through the focus, intention and actions we take. When life hands us unusual circumstances we have an opportunity to discover and experience creation's infinite wisdom.

Exercising More Heart-Talk:

Part of interspecies communication is stretching oneself into new regions, discovering possibilities. Choose to learn something totally new for you – something that threatens your comfort zone. It is a wonderful way to awaken the gifts. If you are facing challenges in your life, open the door to options. Make no absolutes about the way things have to be. Unconditional acceptance of our circumstances allows us to achieve the impossible. We simply move to the next moment without all the ands, ifs and buts.

I'm choosing to learn...

Fear distorts our view
Our interior landscape becomes
a treacherous terrain
Our exterior view
is seen through
these changing rhythms
of our mind.

All life recognizes kin.
All of life is kin.

All we need to do
is acknowledge
all is kin.

Please Spread Those Feathers

Scottie, our precious grand-daughter, was seven and we were taking her to a zoo in Richmond, Virginia, on one of our visits. Scottie and I had a running argument. She thought I was silly believing that animals talk, and I thought she needed help to understand what I meant. We had just finished petting a giraffe when she spotted a peacock. He was at the far end of the very large cage.

"Oh, I've never seen them with their feathers open".

That was enough for grandma. I got her to agree to "send love" with me to the peacock. I followed my usual protocol and when I felt ready the message that went heart to heart was simply, "Would you be so kind as to demonstrate the beauty of your feathers." Now, we had been in this small zoo for over an hour. Although Scottie had not noticed him before, I had. He had remained in the corner all the time.

Almost as soon as the message left my heart, about a minute after Scottie and I beamed love, the peacock turned around, walked over towards us and graced us with his plumage fanned out as wide as he could. He repeated this several times. I could feel his being waiting to know if this was good. I kept sending him back messages of marvel at his being and wonder at his beauty. He then gracefully turned and walked back to his contemplating corner. One peacock made a believer out of one little girl.

Exercising Heart-Talk:

Go talk to a lot of different animals. Stand in their presence as if you are greeting Creation for of course, you are. Mind to mind, heart to heart, send blessings and gratitude for the meeting. Notice each response. Some will be very aware and respond immediately, others will not pay you the slightest attention; some will play shy for a while and others will display fear or anger. Each being responds to the Light from his or her own state of being. Do not take anything personal, not even success. The more you practice, the quicker the centering, the easier the signals go out through you and to whomever you are focused upon. Be very careful at this stage. Focused power is leashed by ethics. Never send anger or fear. Keep them to yourself. It is reasonable to experience an array of emotions, it is not reasonable to wield them as a power against others or back on yourself. Fear can harm yourself as anyone knows. A wise nursing instructor once told me, "Place your negative emotions in a bag when you are entering into your work. Pick them up afterwards–if you still want them."

Heart-Talk Notes

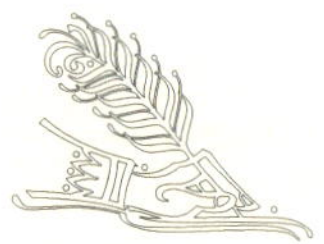

Exercise Healing From The Heart:

The Bronx Zoo in the 1950's was a scary place for me. Watching huge animals pace back and forth in a tiny cage made me cringe. It was easy to imagine how bored and depressed they must be. Every child who has ever been confined to a room for three minutes knows how uncomfortable and scary it can be. Then to be separated from their own community and the familiar terrain of where they came from – it was too much. If I had to be at a zoo because my parents believed I would somehow benefit from the experience, then the animals must benefit too. I stood silently praying for their peace. After a few minutes I would feel their attention turn towards me. I would send them love, heaps and heaps of friendship, telling them I am so very sorry humans did this to them. Before leaving I would tell them I'll always be their friend – call on me when you need to be comforted and somehow I'll be there.

I certainly did not understand spirituality or metaphysics. I understood that humans were cold, and inconsiderate of the animals. Entertainment at a high price. It was the same at the entrance to the circus, where people who were born with very different human appearances were on display. I was embarrassed that I was part of a species who would enjoy mockery more than friendship. I believe children are innocent until taught otherwise. Many children who felt as I did grew up and changed these prison-like zoos into what most of them are today. Animals who are endangered are now being rescued and some of the species are recovering due to the inspired work of so many who care.

A friend of the family recently took us behind the scenes at Sea World where he was one of the curators. Sea turtles and manatees, each with huge bandages and each in their own little pool, were being nurtured and supported. One of Rick's jobs was to release these amazing creatures when they were completely healed. The manatees and the sea turtles were injured by boats. Here were humans on both ends of the spectrum, some causing pain, some helping to heal the injuries.

We can all be healers. All we have to do is consciously focus our intention on being an instrument for a higher purpose. Let yourself be open and receptive to the universal energy moving through you to another in need. The animal can be rushing to cross the street and barely makes it. If it were me, I'd be shaking, my adrenal gland would be wiped out and I would want comfort. Send the squirrel your love, your joy at their crossing safely. Get into it and watch the magic.

If you see an animal injured and wish to help, besides calling the police go to the animal if you can completely immerse yourself in the role of healer. Mother Theresa understood that, so did my sister entering a quarantined room. When your love and caring is far greater than your fear, your fear will depart. Injured animals can be frightening in their extreme vulnerability. So can people. You are the only person who can judge whether this is okay for you to do. If it isn't, then sit wherever you are, open your hands, palm upwards, ask our Creator to accept your offer to be a channel of healing and then be one. Stay with this until help comes. Just like the sea turtle who let someone take them from their home in the waters, any animal in need recognizes help.

My Experiences, My Belief

Plants are fascinating;
Insects are fascinating;
Birds are fascinating;
Mammals are fascinating
What isn't fascinating
Is fear.

Horses Scare Me

The 1940's was a time of ponies in Brooklyn. I vaguely remember getting dressed up in a navy blue hat, navy blue and white checked coat over a navy blue dress, with, of course, navy shoes and white socks. I was ready for the big adventure. Next thing I remember, my father stuck me on a pony. The pony was creamy colored, docile and old. All I remember was having to smile for a picture and being terrified of sitting on this very large beast. That was not only the first time I sat on anything resembling a horse, it was the last.

At age 31, I attended the first psychic reading of my life. Irwin Grief jumped into a past life reading before I could even grasp the concept.

"You were a young woman, or teenager, when you fell off your pony and permanently injured your spine. That is why you have spinal deformities in this life; you carried them forward."

Although I scoffed at his statement and dismissed it, over the years I've had second and third thoughts. During the last 25 years of finding missing animals, including one pony, healing work with horses, explaining to owners/trainers what their horse was saying, I've been offered lots of lessons, even lots of horses. I've turned them all down. Something deep down inside me says this is not for me. Then, another strange piece got added to the pile.

My husband loves anything For Sale, Wholesale, Cheap Today. In the early part of the 90's, he attended military surplus auctions. Two to be exact. One for him and one for my revenge. I was working the morning he left for the first auction. It is my habit to listen to all our messages whenever I first get home. Here's his message that day.

"Sweetheart, you are the proud owner of 128 used bowling pins! And all for $21. Can't wait to see you."

You don't want to know how my mind prepared for this new collection of junk. It wasn't as if he didn't have enough in our dungeon already. The dungeon is my nickname for the "basement that cries for itself". I swear I hear it pleading, "Fix me; clean me." If I can hear birds, I certainly can hear the consciousness of my own home.

The Dickens came home. Yes, they were ridiculous, and yes they stayed far too long. Almost a year later, with the help of a good friend Lee Wilschek, the little critters were transformed into perfect pre-bunny forms for Kindergartners throughout northern New Jersey. To all the parents who received these as gifts, hope you enjoyed your child's creative use of Dickens extravaganza.

Now, let me tell you about my revenge. Many months went by and like all of us, things happened through me, to me and by me. One of them was my body. It went off on its own adventure, tripping up nerves and muscles until I accepted the temporary help of a wheelchair. Today, if I need to do that I would name the chair. I wasn't as creatively open then. Sitting near the kitchen stove pouring boiling water into my waiting coffee press of vanilla, organic, freshly ground coffee, one cup a day treat, the silly arm and hand attached to the teapot weakened and boiling water cascaded down the front of my opened robe. Of course, I had no clothes on – great for total coverage of third degree burns. Ah, but I had a secret weapon, and if any of you don't know of Willard Water, formulated by Dr. John Willard, and sold in health food stores, read the book "Aqua Vitae", the story of Dr. Willard. I immediately sprayed myself with the Willard mix (1 ounce Willard Water to 1 gallon distilled water). I always keep a spray bottle in the kitchen. Besides coating the nerves so there was no pain, healing was complete without scars or skin grafts in about four weeks. I simply kept it wet with the magic water.

During that time Dick did not know what to do for me. Here was his dancing Nancy in a wheelchair with the upper part of my body crispy.

"Would you like to go to a military auction, you've always wanted to see one."

I was there in a flash. If you've never been to one, picture a dirty, all grey in every respect, humongous warehouse with long wooden tables at least 200 feet long. He wheeled me between these rows of tables. Every so often, a paper with a number was tacked near a grouping of stuff. These were the pallets that would be bid on. They supposedly corresponded to the information on the papers they handed out. Once bidding began, these rows were roped off. When a bid was accepted you had to buy it if you ever wanted to come back. That's how Dick inadvertently bought bowling pins instead of some computer supplies. Wrong number.

"These books are fascinating."

A simple thought can lead to lots of work. Eighty four dollars later I was the proud owner of fifteen hundred pounds of the West Point Army Library used book collection. All of it. I thought I was buying one pallet - no - it was the entire row! Our recycling center was kept busy, along with all our friends who walked away from our garage nicknamed West Point #2, with arm loads of books. Then it happened.

While Dick was outside by the open garage door, ripping covers off of books to go to the recycling center, I received a fax. Tearing it off the machine I went to tell Dick. Before I could open my mouth, he spoke.

"Nancy, do you remember the poem; *How Do I Love Thee?* I've got to say it at a wedding this week-end and I'd like your interpretation of the rhythm and feel of it."

"I love it. Look what just came on the fax, the same poem, a couple I'm working with want it. The woman wrote that she is a cousin of Elizabeth Barrett, her last name being Barrett, also. I haven't heard that poem since high school. I really loved it, too. Isn't it strange, both of us get to say it miles apart on the same day?"

The next day my friend, Lee Wilschek, arrived with her 13 year old daughter Adrienne. Lee and I plopped two lawn chairs in front of the garage and continued the process of preparing the books for recycling. It had been a year since my infamous purchase. Many friends had friendly chuckles, but it was time to get rid of all the now moldy books that had served their purpose the day I bought them. Adrienne wanted to start in the back of the garage by the wet, moldy papers. Warning her to be careful and not stay there too long, Lee and I settled back to talk and scan the collection.

"Oh, Elizabeth Barrett Browning."

"Adrienne, what about her?"

"I'll show you."

Among the hundreds of ruined books in the dark and wet garage, Adrienne found something to do with the author of *How Do I Love Thee?*

"Here."

Adrienne handed me a dark blue hard cover book. The spine stated *MRS. BROWNING'S COMPLETE POETICAL WORKS* - Cambridge Edition, Houghton Mifflin Co. The front cover has a gold wreath encircling ELIZABETH BARRETT BROWNING. On the first page a signature states "Erica Clarkson Barrett, Philadelphia, Dec. 13, 1914." Opening this remarkable piece of literature, I discovered that it was the only one of my military auction books still in perfect condition. It also looked as if it had never been read or abused. So I read. And stopped abruptly. Elizabeth Barrett Browning was a happily physical child until 15 years old. She attempted to mount her pony unaided, causing serious spinal injuries which left her an invalid and led to her death in her mid 50's. I started thinking of the weird and only hospitalization I had as a child, at age 15. Nothing ever found, but I lost weight like crazy and was in constant pain.

Years later a spinal tomagram showed that I was born with serious congenital defects in my lower spine that led to my own injuries. I have spent my life learning healing techniques as a result of these problems. I read further. Elizabeth started her writing at age 31. That was when I entered my new vocation and left my traditional nursing behind.

There are no conclusions except to say that the synchronicity tells me there is a connection between that wonderful authoress and myself. She would appreciate my discomfort at climbing onto the back of a horse. She would understand my desire to write thee and thou and speak it when I was about 14, 15 and 16. It felt better than Brooklynese. Whether we knew each other, I was her, or I have access to her energy, her soul, her spirit; something does reach across the boundaries of time and space. Although it is possible to make declarative statements, it seems silly. Every answer I glean is from my own perspective, through my own filters, ideas, and so on. What if when I die and go on I discover that it isn't quite the way I thought it was? So I make no absolutes and no apology for thinking there are alternative answers. What I will tell you are other excursions into synchronicity, the realm of togetherness. Whether I ever ride a horse or not, whether it is my karma, dharma or dogma won't even matter. If I enjoy the connections I have with horses, snakes, cats, then it will surely balance out any problems I have with horses, snakes, dogs and cats. I have to mention snakes; I am in the process of learning to love what was behind my greatest fears. This is a simple belief I hold: all fears of external – outside the self life forms – are illusions. The real fear has been masked by something else, something handy to externalize what is difficult to face internally. The greatest challenges hold the greatest gifts.

Everything we fear represents some aspect of our being; the object of the fear is simply a manifested mirror. Having family members terrified of other living beings actually helped clarify the truth. It couldn't possibly be the spider's fault - no one in my family had suffered at the hands of the insect, cat, dog, or other being. No horror film indoctrinated them; their fear was unfounded. These bugs and animals did not create the problem. Because it was so obvious, it actually served to guide me in my search for understanding.

Today, after years of changes, therapy, meditation, injuries, near death, births and deaths, the tenuous thread is now a strong and loving rope that guides me. Having had a snake phobia most of my life, I know that all the hatred, prejudices, and fears are simply the negative portion of my own personal mind, not about anyone or anything else.

Gathering the best of our thoughts, the sweetest and kindest of our beliefs and manifesting them, guarantees us a moment to moment joy that no fear or hatred can supercede. I believe that is why the poorest of the poor can have a smile that melts an iceberg.

Spirit is fully present and I have seen this joyful manner in Herman the fly, Baby the ground hog and Dick, the husband.

Enhancing the gift of communication:

Treat every living being as if there is a divine presence operating the life force within the form. Approach each sentient being with a consciously open heart and a flowing mind. Imagine your personal mind is one track with an overlapping universal mind track that is not just connected to all else, but is all else. Acknowledge, asking for the universal mind to vibrate at its highest divine resonance as preparation for your conscious connection with other life forms. Allow your higher self to observe the depth of the interaction between you and another.

Some of my experiences in meditative states

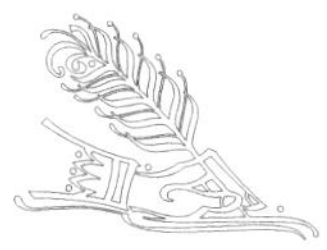

Journal of observations:

My earliest memories of an interest in interspecies communication is ...

After practicing my own centering techniques and sending a universal message of oneness, the difference about this encounter is...

The fool in the tarot says it all; when you are willing to know that love is always the answer, leap into the unknown with absolute faith.

All language is a translation
of feelings and thoughts.
All feelings and thoughts
flow along the infinite waters
of Creation,
retaining the universality
as individualization
weaves its fabric into existence.

It's Okay, She's A Healer

Back in 1976, when I was short and young – just kidding, I was never short but I was younger, I was walking towards an exit in Grand Central Station in New York City. I always enjoy telling my students, "If you can meditate in Grand Central Station, you are centered." A voice screamed out, "Oh my God, I've been bitten." Turning around, I saw a young woman, normally probably pale, go absolutely gray with eyes stuttering in fear. I walked over and she immediately lifted her T-shirt a few inches to show me her very swollen, rapidly reddening and spreading bite by something, on her abdomen. I took her by the hand with, "Come with me; I'm a nurse," and led her to the ladies room. Like magic, a chair was there waiting for her. As she sat down, I let the power of the urge to heal guide my movements, words and actions. One hand on her head, one over her abdomen, I instructed her to breathe in cool, pale blues and greens to the area. Not more than five minutes went by when she opened her eyes and looked down remarking that it no longer hurt. We both saw the area; it was as though nothing was ever there. We held hands walking out to the information booth where two police officers were waiting to talk with her. "Thanks, but I no longer need help. She healed it; she's a nurse". For me, the amazing piece was her trust in the process and her easy acceptance of what to me was a small miracle.

These moments led to a quantum leap in risks. Blurting out I'm a nurse was one thing. Arguing with a trainer, who sees his horse daily and does not see what is wrong, is either ego or totally centered, at least for the moment.

There were four horses I was asked to evaluate for dear friends. They owned them, but like most race horses, they were being trained by experts who also were their keepers. The trainer was also one of the top racing jockeys. He watched, with amusement, as I saw the energy of each horse and spoke of their needs. His puzzled expression turned to annoyance when he discovered that I had never met these horses before. He was determined to unmask me as the charlatan he believed I must be.

George (the jockey) asked to show me another horse and took me to the stall. Standing outside the stall, only seeing the horse's head, I stated, "The horse goes lame in the front left leg on an inside curve, compensation problems probably".

"Wrong, it's the right leg".

Back and forth we went, each asserting our opinions, until I asked to have someone walk the horse around the curve. A groom took the horse to the track. The front left leg began buckling as soon as they walked around a curve. George was now mad.

"One more horse, just this one more."

Now I stood in front of a stall of another horse letting several minutes pass.

"George how am I supposed to tell you anything? I don't speak French and this horse does not know English; it's from France."

"My god, how do you know the horse is from France?"

"She told me."

That was the end of it. I'm sure he now believes that horses speak in languages, but I don't. While centering myself and opening up to the heart-talk between the horse and myself I began to see pictures in my mind of France. It was easy to determine the feel of the images was not common to me, so I assumed they were coming from the chocolate filly standing in front of me. I sent back a thought of home and felt an overwhelming sadness in reply. Back and forth we went until some of the sadness was replaced with our friendship. As I write this the spirit that lived in the body of the horse is helping me tell you what occurred. We have remained friends ever since. Can't say the same for the jockey. He remains afraid.

Not wanting to ride has never limited my desire or ability to be with horses. The chocolate filly is one of many horses I've spoken with over three decades. If I had let my fear of the unknown guide me, or my lack of intellectual knowledge of horses and their lives, I would never have met these wonderful gentle giants.

Years later, Sue Ellen M_________ came to my office. Her bag was laden with objects used by a horse she had recently acquired. Days later I received this email.

This is Sue Ellen's email.

Dear Nancy,

(I had a session with you 3/23 evening and you spoke with Panache by touching his black halter.) Before coming to you, I was intimidated by Panache's size. You told me that my thoughts were smothering the power of my heart. I followed your instructions. After I said hello, I took him out of his stall and stood him in the center of the aisle. I stood by the right side of his head, slightly in front. I took the halter you held and held it gently up against the left side of his face in a ball. "With focused intention," I silently told him that you said hello again and again.

After 7-10 seconds, he let go his head, lowered it and moved his muzzle into my chest and held it there. Then I snuggled him.
We've been friends ever since.
Sue Ellen M_________

Probably the most important lesson I can share with others is this: give time to make adjustments to the exchange of energy information between you and the other being in your midst. Wait to approach physically until you feel the spiritual connection. For me, it is an overwhelming feeling of love that envelopes both of us in a safe zone where we can now be communicating spirit to spirit.

Exercise: It's Okay, I'm A Healer

Create a quiet space; one where you can close your eyes and drift slowly into a relaxed state. While in this centered, deep breathing, relaxed state of mind; create an image, a belief, a thought that you are an ancient and wise healer. You have learned the tradition passed down to you throughout the ages. Your energy becomes magically linked to the energy of the Creator Who Created All Of Life. You are now an instrument of the Creator for peace, for love, for healing. Take any concern or worry you have for another species, whether it is for one specific friend or for one you have heard or read about. Place that particular being in the Light that is in your heart flowing from the Divine to you, and then towards all. Stay there concentrating on that for as long as you can. Usually one to five minutes for first timers. It gets longer as you practice.

Letting the intuition and that which is beyond the self take over, takes practice. During the few minutes when you are simply observing without judgement, you might "receive" some information or feelings. It will be clear they are not your usual thoughts or feelings. You may just have linked with the animal you were concerned with. If you can carefully assess your thoughts, feelings, images, ideas, words that pop up and anything else that occurred, you're on your way to that wonderful world where St. Francis of Assisi entered during his awakening.

My findings

Exercise: It's Okay, I'm A Healer

When my daughter's fourth grade class had a spring break we offered to take home the pair of love birds that lived at the Montessori school she attended. We decided to put their cage in the living room.

Every Thursday night our living room would become the classroom for adults interested in meditation, creative visualization, healing and psychic development. Two Thursday nights passed before the birds were brought back to the Montessori school. There was an egg in the cage when we returned them. The teacher, who had raised these and other love birds was surprised. They had never laid eggs before. We all came to believe it was the loving peace that my students entered into that gave the love birds the atmosphere they sought.

Find the time daily to release the stress, let go of fear and connect to the divine. It's been done forever because it works. No matter what your prior belief system, devote yourself to your own peace and watch the world around you enjoy the flow of light and love that emanates from your presence.

My discoveries

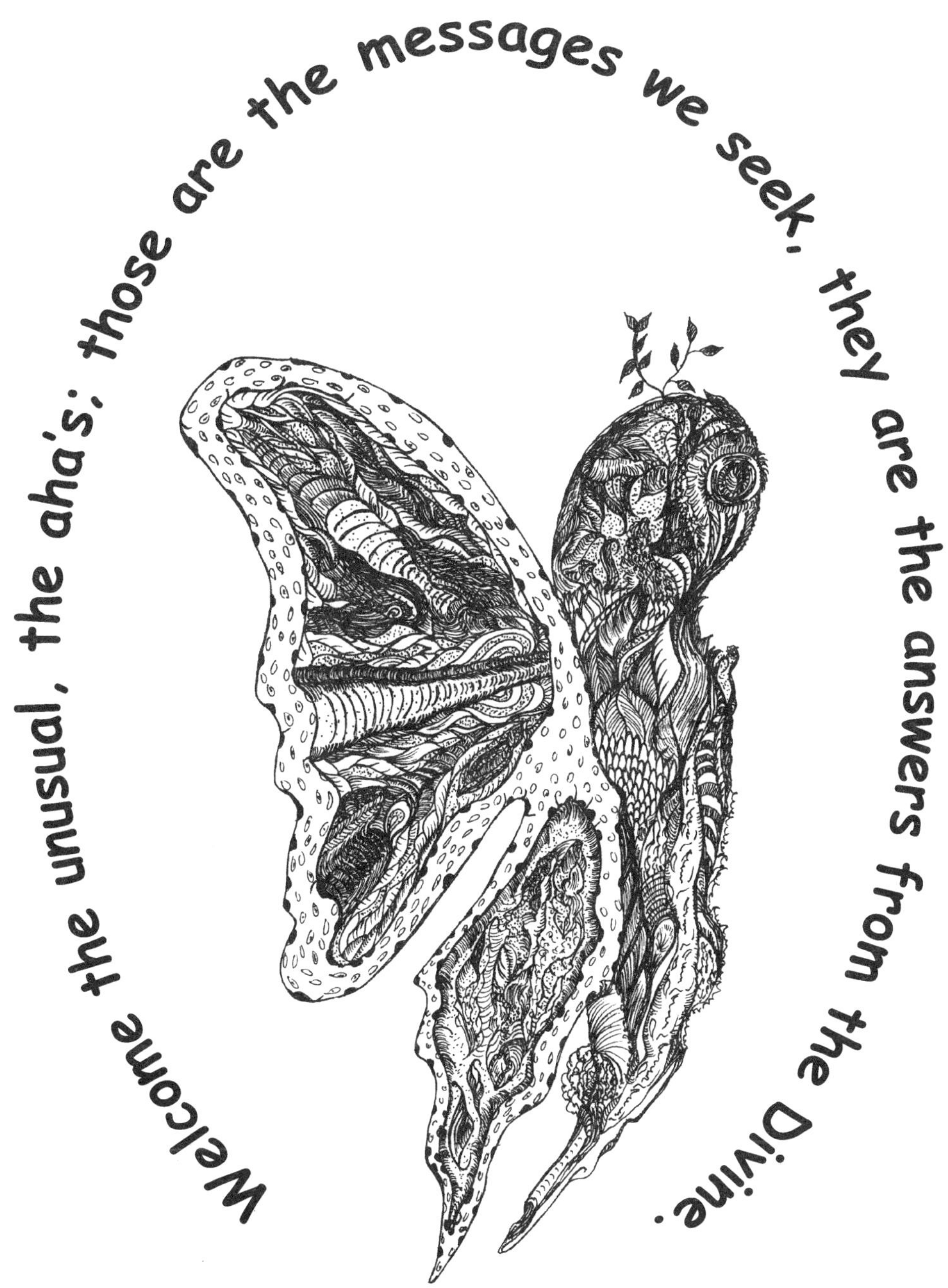
Welcome the unusual, the aha's; those are the messages we seek, they are the answers from the Divine.

We are instruments
of the Light
And so are
all other beings;
Open up
to the source of
all Creation
And all will
feel like family.

Animals Bring Me All Kinds Of Friends

In 1979, I moved to Budd Lake, New Jersey. Rebecca, then 10 years old, and Jesse, age 5, both wanted a dog. Off we went to an animal shelter. I heard from a few people about Alibi Acres. I was told that the woman, Lorna, who owned and ran it, was wonderful with animals and never put them down.

This short, blond woman with a huge grin and blue eyes that danced with delight at our interest in rescue animals showed us around. The dogs were all well taken care of; the cats ran loose in a barn like setting, obviously used to getting lots of attention from Lorna. She brought out a mutt with long brown hair, deep sad eyes and the sweetest disposition. This little four legged girl came over to us with fear spilling out of her eyes. Her movements showed that her body was hurting.

"She was abused and injured, she now has seizures."

That's all we had to here. She was ours. We named her Ramona after a song from one of those black and white movies I love.

Marianne was a journalist and a Society For the Prevention of Cruelty to Animals agent. Our first meeting occurred in the early 1980's.

Marianne was on the animal welfare board of Alibi Acres that Lorna founded. The three of us, Marianne, Lorna and I, had all been working with horses in one way or another. At the time, Marianne was writing for a statewide horse publication as a reporter. Marianne wanted to interview me for the New Jersey Horseman. She wanted the interview to be from the animal's point of view, which she hoped I could provide.

The horse she chose for the interview was Lad. Marianne had a special affinity for Lad whose early history was filled with painful abuse. Lad had been with Marianne for about a year when I read him. I felt an instant and easy communication, as if my mind and his spoke the same language. No need to spend any time interpreting my sensations or feelings, just flowing thoughts and words. Lad "told me" he would live to be twenty-eight to thirty years old. Last I spoke with Marianne, he was twenty seven and in excellent health. I saw him winning ribbons, dozens of them; he did, indeed, go on to win every award he was eligible to win. Another piece of information Lad told me was about a piece of broken cartilage on the end of his nose. This had been broken when his previous owners had twisted it while abusing him. No one knew this at the time. Marianne had a vet check it and he confirmed that it had been broken. It was so far up in his nose that without knowing it was there, you would never have thought to look for it.

Interestingly, a few years later, Lad competed against a horse which belonged to the man whom Marianne had bought Lad from–the man who had been Lad's abuser. Marianne told Lad, "You know what you're going to do? You're going to beat that S.O.B." And so he did. Lad had been 200 pounds underweight due to neglect and physical abuse and terrified of men when he first was sold to Marianne. Her love broke the circle of hate/fear and wove magic in this now outstanding champion.

Marianne's rapport with Lad became the zen experience of being one with each other. She could hear him talking in her head. She had conversations with him all the time and knew what he was going to do before he did it. He became spectacular in the show ring.

Maniac was Marianne's Himalayan, longhair cat, and at the time he was, according to her, "The biggest slob that ever walked." He would not keep himself clean and he hated to be bathed. He had a coat that looked like "who did it and ran."

Underneath the mess I saw a winner. After a moment's introduction to Maniac, I told Marianne, "Someday you are going to show that cat." She laughed. Seven years later, when Maniac was twelve years old, he became Double Supreme National Household Pet. He went into the show ring for the first time at nine years of age. Gradually, he began to groom himself impeccably and become very proud of his appearance.

Because of my friendship with Marianne, I volunteered to help on several cases of animal abuse. In the mid 1980's, the SPCA agents of Morris County were dealing with extreme cases of animal cruelty such as ritual killings and cockfights. Then there was the case that made headlines.

The neighbors of a family that had two great Danes complained that the dogs had disappeared, but their collars and chains were still there. They accused the family of killing their dogs. Marianne called me. I told her their son did it, and they were buried in the back yard in plastic bags. I warned her to watch out for him, as he was very disturbed and dangerous. "The mother will answer the door, and the son will be listening from a flight of stairs with a shotgun", was the image I conveyed. Marianne took my advice and went in with a police officer.

When they arrived at the house the mother answered the door. The son was listening from the top of a flight of stairs, and he was holding a shotgun. A call went out for more armed police officers and they went in. The police and Marianne found the dogs buried in plastic bags in the yard, killed by a shotgun.

The mother knew her son had shot the dogs, but did not report it because she was afraid of him, with good reason apparently. He was prosecuted, and in the end, the court ordered psychiatric help for the boy. The probable cause for the boy's behavior, it was found, was severe child abuse by his father early in life.

Yesterday, June 27th, 2002, I met with two wonderfully wise souls; one happens to have the body of a dog and the other, the body of a human. This story shows the capacity of a soul. Pat is a friend of my husbands, BN (before Nancy), and remains a good friend who we both enjoy visiting. My husband Dick, had seen Pat the previous week and suggested I might want to meet his new dog Innish.

We parked our car right outside the glass door entrance to Pat's business. Before getting out I wanted to ease the tension between us. We were having a discussion and were both upset over problems one of our children was having.

"We need to be connected, flowing with the energy. Let's find a way." That was my need speaking. I knew meeting a dog who might be sensitive to everyone's energy, it was important not to carry yukky stuff.

A beautiful black and white Springer Spaniel sat looking at us from inside the door. As we opened it he began a low growl. Pat looked at Innish and said, "That's odd; he didn't growl at Dick last week."

I responded, "He wouldn't, he's not afraid of men, only women – women who hold any tension."

"Yes, that's true. He growls at my daughter and she loves him."

"But she's having a difficult time right now that is causing her some stress."

"Yes, she is."

"Okay, when the problem is gone and she learns to clear her stress rather than carry it, Innish will feel safe."

We then proceeded to a room where we could all sit. Innish did not leave Pat's side and Pat did not stop peacefully stroking him.

"I see a woman with a stick, she's lifting it up to hit. Innish received lots and lots of beatings, starved a lot, and, let's see, never was allowed in the house?"

"It was a shovel that she hit Innish and his brothers with over the head, again and again. They were chained up outside for breeding purposes only. The neighbor who called me told me the dogs were never let in the house. The owners of the house went away during the winter and left the dogs to starve. Everyone was afraid of the owners who would suddenly pop up to apparently get something inside the house. Two of the dogs died. The neighbor finally called me. I went every day of the week for one week. I sat opposite him most of the day. By the sixth day, I attempted to get Innish to go with me. He's incredibly bright. I've never had to housebreak him; he automatically understood. I can tell him to stay; I'll be right back, and he does."

Innish is incredibly bright in many ways. Pat never changes his inflection or tone, no hand signals, just conversation with a dog that completely understands him.

I laid down near Innish and peacefully centered myself. I could feel Innish's past tensions slipping away, as I did Innish took my hand with his paw and motioned for me to put my hand under his head. He peacefully closed his eyes as he laid on my hand.

The remarkable wisdom Innish has comes through clearly. He is a great teacher, showing how to love and trust in spite of abuse issues. His energy and health is wonderful despite his past. He has allowed himself to recover.

Exercise: Illusions and Truths

Infinite intelligence takes the form of a human, a cat, a dog, a fly, a bird. That is my recall of a passage in *Kinship With All Life.* Write your own experience of discovering some other species ability to show you that you are not alone in your intelligence and wisdom.

Exercise: Practicing Intention

It is always easier to change our routines and habits when we are not desperate to change them. If you are not in the midst of overwhelming crisis, find the time each day to practice intentional sending and receiving of love. If you are in the midst of too much tension to even find the time, then select a coach, a mentor, someone to whom you will listen, to guide you inside. At the end of the book is a page of selected tapes and Cd's that can support your desire to become an animal communicator.

Now, for those who are ready, here is a simple exercise. Place yourself in a position where you can be comfortable, without restricting your circulation and flow of oxygen. That means either lay down, without crossing ankles, legs, or arms. Place pillows wherever you need to relieve tension, or sit with your back straight, head centered (please look in a mirror once to check yourself out) and again keep arms and legs uncrossed. Loosen all restrictive clothing. Relax and deep breath. With every breath you take, send a command from your mind to your body, instructing your body to let go of tension. In a tape I produced, World Within, I walk you through this in detail using SuperLearning music to speed up the process.

Once you feel you are as relaxed as you can be, "notify" your heart center by breathing in trusting, loving thoughts like: "my heart is flowing with universal love and spirit". Imagine or actually see (we'll get into that later) a beautiful light stream of energy pouring in and out of the center of your chest. Imagine the universe is returning the love to you, tenfold. In Eastern Indian mysticism, it is considered the Heart Chakra. Chakra means "spinning wheels of light" and is visible to those of us who are clairvoyant. I can tell you when someone is sending love; it is as beautiful as a Michelangelo. And when they are sending hatred, it is as frightening as a war movie.

When you are comfortable with learning to relax, you can be standing in the middle of Grand Central Station or facing a crazed, raging beast (or person). It doesn't matter where or what, just keep centering yourself. When you feel you are calm and fairly emotionally unattached to the outcome, send your first message to the other life form.

Here is what I usually send: "From my soul, my heart, to yours, through the light of the one who created us all. I offer you my love, strength and joy at knowing of your existence. I wish you good health and a long life. Please call on me for anyway I may be of service. Thank you for letting me join with you for the moment."

Some of the greatest teachers are found in our so-called pets, souls who have willingly cast aside their solitary exploration to restore trust and love in a human.

Taking A Different Step

While camping out in the west, around Utah, I met 2 young men who were also camping. They had just encountered their first rattlesnake. Tim had gone up the trail first and half way up, about six feet in front of him, mad as hell, was a rattler, rattling away. Tim's heart raced faster standing still than it ever had running. He decided he had no chance so he might as well gamble on becoming friends. This all happened in about 2 minutes. Let me also explain that while Tim did not know rattlers well, he was intelligent, knew something about snakes, could hear the rattle and saw a striking pose. Instead of running, Tim kneeled down, "to make myself non-threatening," and spoke to the snake calling him brother and friend. He said that all the while he was talking he was sending pictures of what he believed was friendship. He couldn't be sure of how long, and it did feel like hours, but his buddy said it was only a few minutes later that the rattler bid them farewell and left them alone on the trail. What about these wonderful, heroic animals who do all kinds of rescue, even at the cost of their own lives? Certainly they can sense our needs. We need to sense theirs.

The lesson Tim taught me was wonderful and simple: When fear does not guide us, hearts can lead us home.

Dr. Ruth Jones Remembers

"We were at Mike's stable and there was a little pony, a Hackney (the kind that pulls carts) with a lot of knee action. World champion, but wouldn't move right. Nancy looked at it and asked, "Where is your foot most comfortable?" He picked up a foot that had been pointing straight and turned it 45 degrees laterally. Then Buck knew how to shoe him and he was fine after that.

Then there was Mr. Majesty, the race horse that Ronnie and Pete were going to sell for meat. Ronnie met Mike and convinced him to take Mr. Majesty. Before Mike could finish the deal, he agreed to let me call Nancy. We spoke on the phone and Nancy said to look at his kidneys. It was found that Mr. Majesty had a serious kidney infection and that's why he couldn't do anything. There were no overt symptoms and the several veterinarians called in to look at Mr. Majesty had not considered a serious infection. After treatment he went on to win a handicap series at Penn National.

Watching Nancy easily slip into knowing horses I have loved, worked with, own, and been friends with, rekindled my interest in the healing arts. A few years later, I pursued my dream of being an Osteopath. Today, the Wellness Center I operate in Bedford, Pennsylvania, has incorporated my own devotion to the raising of awareness, the fine tuning of spiritual intention and intuition, and an occasional call to a dear friend who can sometimes find out what is going on with my human clients by simply raising her level of awareness to include the entire universe."

Exercising Our Mind's Eye:

The evolution of trusting our intuition, no matter how psychic we are, can be a difficult task. Anytime we have been scolded, ridiculed or ignored as a result of attempting to apply an idea, we may have absorbed a message that our intuitive concepts get us in trouble. Enough of that and we have to undo the problem, recover the trust and use of our intuition. What happened for me with the French horse and trainer was the perfect mixture of antagonism, happiness at a new discovery and rebelliousness (mine, in case that's not obvious) and a tiny drop of annoyance. Stirred together, I was able to make the leap into feeling more confident of the "knowing".

Using your imagination, or clairvoyantly seeing what occurs, "see" the energy of your prayers, wishes, and thoughts flow outward toward the heart or imagined heart of the other being. It does not matter how large or small the being is, nor what mood in which you assess them.

Creativity gets to play here. Depending on what you want to happen, you might send a request, **never** a command. Sit back, relax, enjoy and visualize!

Exercising Our Mind's Eye Again:

Continuing this exercise, gather photos of animals you love. They may no longer be on earth, or they may be lying at your feet. Sit quietly and when you know that you are as relaxed as you can get, hold one of the photos in your hands. Let yourself float gently, as if you are gliding on a lake that glimmers with placidity.

Converse with the pet in the photo. Give yourself plenty of time to feel or hear a response. Rushing through this exercise closes the door to the information. Fear of what you may or may not find acts to create a chemistry that closes off the information. After a few minutes, open your eyes and stay quiet. Mentally converse with the friend in the photo. Imagination is the main ingredient to entering into the state necessary for success. Trust yourself. Give yourself lots of these exercises before you place judgement on the outcomes. When you feel like exploring further ask your friends for photos. One cautionary note. It is important to believe in yourself and also to be graciously tactful. If you believe you are picking up information that is frightening or uncomfortable, take it inside. Send the information to the Divine Light, ask God for the right words, the right tone, the right way of handling what you think you know.

Conversations with my friends

Nancy, Amir and Linda Hepburn
at the first of many meetings

Melting Fears

Several encounters with pain-ridden, frightened, angry animals have taught me that coming from the heart not only makes it safe but can work miracles of healing. The first example I remember spotlights a shaggy dog.

My girlfriend, Phyllis, and I were planning to go out for lunch. I had just walked into her home; we hugged and got ready to go out when she asked if I would mind helping a friend of hers.

"Her dog has a problem. We pass their home on the way to the restaurant, so can we stop there?"

"Fine, just remember that I don't want to know what the problem is."

We pulled into a driveway. Her friend was waiting for us at her door. She opened it and out bounded this 100 or so pounds of shaggy fur. I decided to bound along with the dog. I quickly ran out to greet her or him (remember I only knew it was a dog). We rolled on the grass for a few minutes and then happily collapsed together, dog in lap, my face in her fur.

There was absolute silence around us.

"How did you do that?"

"Do what?"

"Get Jasmine to play?"

"What's so hard?"

"Well, we needed your help because she attacks everyone but me!"

"She's just afraid of how people handle her. She's has a lot of joint pain. Hasn't the vet talked with you?"

"Yes, in fact we were told she's got arthritis."

"Well, if your joints were as swollen and painful as hers, you'd be scared of being touched except by those you either know well and are accustomed to or those who know how to touch you."

The rest of the meeting was spent discussing ways to assist Jasmine. Homeopathic remedies, herbs, nutrients, diet and, above all, a healing touch were some of the avenues we explored. When we left, Jasmine tried to join us. Knowing the restaurant would not accommodate her, we kissed good-bye and we both trotted off to our individual worlds.

In the same year, a woman named Linda called. From the moment I heard her sweet, open, direct, and caring voice, I couldn't wait to meet her.

"I read an article about your work with animals. I have a horse that I need help with; would you please come see him?"

"Sure. What day?" It didn't matter that the last and only horse I had been near was actually not a horse; it was a pony in Brooklyn. The photo of me on top of the pony and my sister, Anita, standing by our side proves that I once was near a hoofed animal who had a mane.

The day was the following Monday. Linda lived in what is commonly called God's Country – acres of flat land, flowers by the side of the road, horses grazing everywhere. Looking past the barns and homes, the land rose like a tidal wave, covered with froth made of trees and stone. The mountains were red and gold with touches of green.

A woman, who looked like a gazelle with long straight light brown hair, was standing in the driveway. She had stepped out of a back door when my car pulled up. Linda had a grace about her as if she were flying free across the land.

"Nancy?"

"Hi Linda."

"Would you mind if another person watched besides me?"

"Not at all."

She went in to return a moment later with a man carrying a camera.

"Hi, I'm Robert. I work at the local paper. Linda called and told me you were coming out here. Would you mind if I report on what I observe?"

"Not at all."

I followed Linda to the barn.

"Remember, I don't want to know anything about the horse, except perhaps its name. I like to call everybody by their name."

"Amir"

"Thanks."

We walked into the barn. A horse called out. Linda walked over to a stall.

"Amir, this is the lady I called to help."

Her hand was stroking his face. Linda put a rope on him with a very short lead.

"Would you mind leading him out to the field with Robert? I have to check something."

Not bothering to tell Linda I had never been near a horse, I said, "Sure," as I took the lead and started chatting with Amir and Robert as we walked.

"Hi Amir, come on sweetie; let's go to the beautiful fields."

We stood in the middle of a field scattered with bright colors. Never having learned the names of most flowers, all I could spot were the daisies. The blue flowers were like stars that dropped from the sky; pink ones looked like delicate air brushed fluff and yellow ones looked like pieces of the sun's rays. It was beautiful and helped me center my mind on Amir.

Connecting with him was easy. I soon found a feeling that translated as a history of trauma. My body seemed to change into a horse's, except that I was the only one who knew that. My front shoulder ached; my head had old damage from a fall as a baby. I could see me as a young colt falling backwards. Then, the image changed to one where all my insides grew sluggish and tired. I knew that feeling well; I label that one allergies.

Translating my experience to Linda and Robert was easy. I simply deleted the part where I became an animal. We kinetically muscle tested Amir for the feed he ingested. I used Linda as a proxy. Two products seemed to show a weakening in the system. Linda decided to switch to other feed and see if that would help.

"I want us to win in the shows. I know he is good enough."

"Linda, he gets nervous when he doesn't know what's coming. Not unlike many of us, he handles change poorly. If you can go to the place where the show is being held beforehand and firmly fix an image of it in your mind, it would help. Take that image and stand before him, imagine making mental contact, third eye to third eye, and send the image again and again along an imaginary beam."

At the conclusion, Linda again handed me the rope. Casually, I took it and pulled Amir along. When he resisted, I tugged laughing, "Stubborn as any man I've ever met." I kept on pulling and teasing until I turned the rope back to Linda and started to breathe again. Glad that he was fun and I could handle him I said, "He was an absolute joy to work with."

"Yes, he was quite different today, thank God. Usually, he kicks and bites. That was one of the reasons I called you. Funny how good he was; you sure do know horses."

Back in the comfort of my car, my mind shook harder than my body at the risk I had just taken. Sometimes I get in trouble and sometimes I don't, but I always like to jump into the world while it's spinning.

Linda called a month later to tell me she did everything I showed her and they won their first prize easily. In the years that followed they consistently won and grew very close as friends.

Linda invited me to her wedding two years later. When she and her husband separated, she came to live in my home in Budd Lake. Amir was moved to a friend's farm. Six months later, Linda was ready to move to her own place.

Linda attended my wedding on February 11, 1989. A month later, her sister called to tell me that Linda went mountain climbing for the first time. (She went with her boyfriend, an avid climber, and his friend, an expert teacher.) Linda fell off the mountain and died. There is so much more I want to see happen for her that never can be and yet I know Linda is handling her world better than her family and friends are handling missing her. As much as I understand and believe in life after life and reincarnation, it doesn't make the loss go away. It does add other ingredients: comfort to know she's okay, mysterious meetings out of body, her voice ringing true and clear in my mind. All these stretch the emotions into acceptance.

By 1984 I had at least four years of working regularly with animals. That year, a call came from Jim, a trainer I had met through a wonderful couple, Marlene and Galen. I had worked with them as individuals and as pony trainers. They are two of the more dedicated and caring animal owners I have ever run across. Galen has gone on to teach many other people healing techniques to use on people and animals and, when last we spoke, was still thick in the study of healing modalities.

It was a beautiful autumn morning, a perfect day for a long drive into Pennsylvania horse country. The drive made the day more like a holiday in the country, not work. The roads were lined with blazing fall colors everywhere, in the sky, on the ground and in the process of leaving their temporary homes on the ends of the arms of their parent toward the earth where they would rejoin with all. It was an awesome journey and I loved watching it happen. The directions were easy, hardly any turns until I got to their town. Two hours of solitude and I was ready for people.

Jim came striding to the car with a gait created from thousands of hours on horses. He had long lean muscles; hair that never looked combed and eyes that showed he had one focus and one focus only – horses. In all the times I spent working with Jim, I never got to know him except for his concern for the animals in his care.

"Hi Jim, thanks for the directions. They were perfect."

As I pulled up, Jim was all questions and nervousness. "How many people can follow you about? Are the owners permitted to be there? We've got about eleven horses for you to see."

"Jim, as long as everyone can fit in the barn, it doesn't matter to me who or how many people show up. Just please tell everyone not to discuss the animals out loud until I've worked on them."

When I am on an assignment for an animal, I ask the owner and trainers not to fill me in on why they called me until I ask. First, I need to connect with the animal, soul to soul, establishing a strong bond between us so that the truth of where we each come from is understood by both of us.

They were ready for me. There must have been two dozen people, a half dozen dogs and unknown number of cats all curious about the newcomer. A dog ran up to greet me. I love animals; they have a way of making me feel welcome, as if they can still feel the shy little girl present in me.

"Let me introduce you to everyone."

Introductions done, a glass of water in hand, someone carrying a tape recorder started us down towards the first stall Jim had chosen for me to enter.

In the stall was a big, beautiful chestnut male with a star on his forehead. He was taller than average and more muscular. The name on the stall read Flying King.

"Hello," my mind sent. "I'm here to be with you anyway that will help. Please show me what you need. I love you."

My head went reeling with pain. I almost fell over in the grip of shooting, knife like pain that was all over my head. Migraines! My God, how horrible. This horse had gripping waves of head pain that kept him both furious and scared. Tears welled up in me. How could he not be crazy, that poor soul. My hands became charged with my feelings of concern. Every cell of the me I knew and parts I was only vaguely aware of became dedicated to helping him be free of his misery. "God, please bring through me whatever powers of healing and love that will aid in the most loving way possible, this soul who waits for help."

My hands felt warmer than they had a moment ago; my fingers were tingling. Gently I placed my hands on the left side of his back. Moving my hands slowly, I would listen carefully for what they could tell me. Images of a network of energy floated before me. Like a greyish blue gauze that has depth and motion to it, I watched it vibrate beneath my hands. Wherever it showed itself to be twisted or I heard a discordant rhythm to the energy, I would leave my hands over the sight and pray that God's Light would pass through me into my friend.

About a half hour later, my hands were massaging and gently pressing the top of his head. His eyes were droopy and relaxed. I could feel a realignment take place in my image of his cranial sutures. My hands walked down to the right jaw, massaging with my fingers, more and more vigorously; then, his left jaw. He then rested his head on my shoulder, gently nudging me. "I love you too, you sweetie." We were in complete empathy, mutually open and trusting. Powerful rushes of his love poured through me as we kept the cycle going.

Not moving from our circle of light, I asked, "Can I have a brush, please?"

A moment later a brush appeared from the crowd. My desire to brush his mane sent shivers down my right arm into my fingertips. As I held the brush, I lost all sense of separation. My fingers, the brush and the mane were joined as one. Long strokes stirred with gentle, coaxing words flowed between us. The brush became a bridge of love, a conduit between horse and woman. His eyes followed me with curiosity.

"I'm just brushing your beautiful mane. I hope you're feeling better."

Gradually I looked up and noticed no one had moved. I was so engrossed in our reverie I could not discern their expressions, only that they looked strange. I wondered what was wrong.

"Anybody care to say something, ask something?"

Jim was the first, "How did you do that?"

"Do what?"

Jim continued, "No one has been able to touch his head since we've had him. All he would do was bite and kick. Couldn't even brush his mane."

"How long is that?"

"Three years."

"I guess if I had the equivalent of a migraine for three years I'd be pretty nasty. Matter of fact, when I have chronic pain there is a particular word most people could use to describe me. It rhymes with witch."

The rest of the day was simple. Flying King had helped me show the people what I do. They were easier to work with after that. I worked on ten more horses, none as troubled as King. Everyone was eager to help their own horse. One woman asked if I would look at a dog, it was having some problems. When people are open to an idea, more creative flow occurs and everything gets easier. Of course, there is a danger to that. It all depends on what kind of idea you are working on. Ethics need to get mixed into the recipe.

When we all said good-bye, I ran back to kiss King one more time and to thank him for letting me be a part of his life.

A few days later, the trainer called to tell me he could brush the horse's mane for the first time. Weeks later, he told me the horse had become increasingly affectionate and easy to work with.

A strange twist occurred when a reporter from a horse magazine called me and wanted references. I gave her a few names to call, among them Flying King's owner. The owner refused to speak telling the reporter that she would not want it known that her horse ever had a problem as the price for resale could go down. I always felt sad knowing that it was more important to keep high money value on an animal than to care for them effectively.

Exercising Healing Feelings:

This exercise brings greater understanding of the healing gifts. Over the last twenty five years of teaching I have watched beginners respond to this exercise with comments such as: "I didn't feel anything, I'm lost", "My hands are hot and tingly; it was amazing", "I'm not sure; it felt nice, but the person I was working with said her shoulder was in pain before we started and now the pain is gone."

Get ready to leave yourself open to the mysterious universe of healing. A mixture of prayerful concentration on your own connection to the divine flow of healing energy and your desire to be of service works well for this exercise. In the book ***"Healers and***

Healing", a common thread is shown. Each of the well known healers of our times states in their own way that it isn't their need to help someone else as much as their need to stay connected to the oneness.

When my sister Anita was 14 years of age, she lied about her age to become a Nurse's Aide. She came home one day telling us about one patient. "She's been quarantined with suspected Bubonic Plague of all things. None of the nurses will go into the poor women's room. So I did. What's the big deal? She doesn't have it. She's so alone; so, I spent a good deal of my time with her."

My mother was hysterical, worried about her baby. Anita ***knew*** with the strongest belief I had ever seen in her that nothing would happen. Nothing did. Days later, a diagnosis of a non contagious disease allowed the woman's quarantine to be removed.

Now, that doesn't work for everyone, of course. The odd thing about that episode was that for most of Anita's early life her immune system was not effective. She had all the childhood diseases including polio at age 11, scarlet fever before that, rheumatic bruises and so on. Yet, faced with apparent dangers to her health, she entered that room. A mixture of utter faith in her intuition and a love of helping caused fear to be so diluted that it was almost non existent.

A cautionary note: it is best to begin at the beginning. Just like a beginning medical student, taking on more than you can emotionally handle would not be wise. Trying to prove oneself dismisses the power of healing. Fear also dismisses the power and a lack of being lovingly open to the divine curtails reaching the objective.

Find either a plant, a furry or feathered friend, or possibly, a human who would enjoy the experiment with you. Your job is to concentrate on being available and vulnerable to the universal healing energies. Openly receive these energies and share them with the other life form by placing your hands within one to two inches (please approximate, rulers spoil the thought) from the edge of the other's body. Keeping your eyes partially closed, focus on the Creator's love for life and let it course through you. Imagine the light of love pouring through you like waves. Keeping your hands within the aura (energy fields) of your partner, imagine the flow of healing energy leaving your hands, flowing into the one you are working with, and continuing on like a fountain of light through that being and back to you. Encircling both of you with the love of the healing energies. Keep this up until you simply feel that is enough. If you are listening to your hands and your heart, the time will vary according to the needs of the other and your ability to sustain a centered state.

Want to increase your focus and healing powers? Take a course in yoga (any kind that interests you), Tai Chi, Chi Gong (spelled a variety of ways), Reiki, Touch For Health, Mari-el, Applied Kinesiology (usually reserved for licensed practitioners), and/or Aromatherapy. Of course, there are others, but the first pre-requisite is to meditate and pray. Tai Chi and Yoga offer us an ability to enter into these states naturally. Years ago, I produced meditation tapes which I still use along with others I have since found. Sometimes I use CD/tape music and meditation to help still my busy mind; other times a walk in the woods is all I need. Find your own way to center and keep at it. This is your best tool.

The more you practice hands on healing, the more available and sensitive you become to the other energy fields. Keep a log of your experiences. Trust any thoughts, words that pop up in your mind and sensations you experiences. Jot them down. It takes a while to recognize the subtle messages, particularly with all our expectations and brain jams on *how could a dog possibly tell me something in my language?* ***They don't***. What an amazing and complex universe this is. Think about how your body translates information constantly from food particles and air temperatures. Does your body talk? In English or French? Perhaps there is a language that is truly universal, a language not of the tongue but of the energy fields that surround and permeate all of life. The energy is like a transformer with the junctions being each individual cell of existence. Right, it's happening all the time, non stop. Now, how easy is it to understand a parrot or a cat or an elephant? If billions of bits of information are constantly streaming through us, the trick is not to learn to tune in but to learn to quiet down enough to focus on one tiny bit amongst the billions. Now you're getting it. Patience, your furry friend would love for you to truly understand what would help improve the relationship between the two of you, or what they need to help them feel healthier. Once you start tuning in, funny things can happen. Suddenly you decide to read about nutrition and pets, homeopathy, aromatherapy, reiki, shiatsu and on and on. Who is the teacher and who is the student?

Taking that leap of faith and believing that your wondrous self can automatically translate the information coming from one specific individualized field of energy know as Fido or Quinnie takes repeated suspension of doubts and fears. It is absolutely worth it. When you suddenly think, *feed her a carrot, how strange, why would I feed her a carrot* - **go and feed her the carrot.** When your dog eagerly chomps down, wags her tail and finishes the carrot, stop and realize ***you just received a clear message and responded***. Note the milestone. Now, there will be many more, so many that one day you will look back and vaguely recall how it was not to feel connected to all other sentient life.

Milestones in trusting my intuition

Exercise: Trusting your intuition

"Hi, what a nice dog. How old is she?"

Ever say that? Next time you are out somewhere and see a dog you like, ask the dog – mentally and quietly of course – "How old are you?" A number will pop up in your head. Now you can ask the person walking with dog "How old is she?" Or if you want, do what I do.

"Hi, your dog is beautiful. Is she four?"

"No, she's three."

"Are you sure? She told me she was four."

"Hmm, let's see. She was born in...right, of course, she just turned four. Yes, she is beautiful. Have a great day."

Most people don't notice or are too polite to ask how I know. Try it. You might have lots of fun and get to know some great animals."

My excursion into the unknown 

Exercise on intuition continues

Every year for the last nine years I have attended and given workshops for the International Women's Writing Guild's Annual Conference. It is held in Saratoga Springs, New York, at Skidmore College. While wandering on one of the campus paths, my girlfriend Jude and I saw a beautiful big Labrador Retriever sitting near a man. The man was seated on a bench holding the dog's leash.

I walked over and kept up a running mental message to the dog. *"I don't know why you are anxious or uncomfortable. Can I help in any way?"*

As soon as I approached, the dog lay down and held up one leg.

"Tommy, what are you doing?" the owner wanted to know.

"I think he just wants me to know where his problem is."

"He's going for surgery on his leg in two days. That's odd, why would he show it to you?"

"I work with animals. Maybe he knows that."

"Yes, I suppose, but it is so odd that he is still showing you his leg."

"He wants me to put my hands on him. Would that be all right with you."

"Sure."

With my hands I touched his injured area gently. With my soul I prayed for a successful outcome. When I felt "done" I stood up. Tommy regained his seated posture, licked my hand and put his head under my palm. I continued to send him positive thoughts and promised him my spirit would be with him throughout the next few days. He wagged his tail.

I've noticed throughout the years that if you come from love and sincerity most people will not question your motives. They will feel your integrity and allow you to move into their lives for a while. All the fears of what others may think of how weird we appear are simply our own insecurities talking to us. Over the years mine have diminished to a whimper.

My healing experiences with animals

Exercise: Intuitive Risk Taking

Terry and Casey were inseparable. Casey was dying. Terry stayed home from work to minister to her beloved friend who happened to be a dog. Terry asked me to take a look at Casey.

Casey was an incredibly easy conversationalist despite his ebbing energy. He lay on a couch with Terry at his side. Casey and I conversed.

"I'm not in any pain. I'm leaving within days."

"Is there anything I can do for you Casey?"

"Tell mommy I was with her before and I'll be with her again. The dog before me was not as close to her. Mommy and I have been together several times. We have grown so close."

"Terry, Casey claims not to be in pain, is at most days from leaving this earth and is quite ready. Did you have a dog before Casey? Casey says you were not as close to the other."

"That would be Charlie. No, Casey is special, very special. He's my best friend. We have an incredible bond."

"Tell her I love the photos she took of me. Particularly how I got ready for them."

When I repeated the statement Terry laughed, got up and walked over to a mantle where some photos stood.

"I dressed Casey up in sunglasses, bandanas, hats and other props. Casey always looked forward to it."

When I left it was clear that although the journey without Casey's presence would be hard, Terry had an amazing relationship that would live in her heart forever.

If someone you know is worried about a pet, leap in with your heart and soul. Follow your instincts. Trust yourself, trust the wisdom that pours through all of life. Perhaps your desire to be of service will lead you to discover how powerful your own gifts are. Although it can be painfully sad, there is nothing finer than befriending good, loving folks and their buddies.

My experience in using intuition to befriend others

Percy The Dolphin

Percy the Dolphin comes to the bay
A truly gifted spirit reveals a special way
He gave our world a message early one day
And here's a vision he asked me to relay

You are young inhabitants on your Mother Earth
Our matriarchal waters we have thanked since birth
Many other species we've watched coming through
Removing one another they also thought they knew

Percy the Dolphin came to the coast of Wales
To guide us with a light, to share an ancient tale
The story that he tells us, the truth that he unveils
Brings to us a heritage proclaiming to prevail

Percy the Dolphin sends us all his love
Universal energies below and from above
Says to know that mother watches patiently
And waits for her children to live peacefully

You are young inhabitants on your Mother Earth
Our matriarchal waters we have thanked since birth
Be assured our planet has and will endure
Eons old the Alchemist, her light is truly pure
Percy the Dolphin sends us all his love.

From a collection "Songs of the Spirit" co-written by
Elaine Silver & Nancy O. Weber - copyright 1987

BUNNY'S MISSING

The woman's voice was spilling frantically through the wire as she implored me to find her dog. Bunny was possibly stolen and she was desperate.

"What is your name?"

"Sue Parks. I'm sorry but I'm absolutely desperate. Bunny means everything to me. I can't get around well. It's hard to go looking. I found your name in Grit magazine. You do find missing pets, don't you?"

"Sometimes, I'll do whatever I can to help."

My hand still held the telephone and my feet still touched the floor; something I call my being or essence was no longer in Budd Lake, New Jersey. I was traveling with the winds but faster than I could be aware of. Seconds ago, Sue asked a question and now I was staring out through the eyes of a beautiful, sweet German Shepherd. She reminded me of the first dog I had as an adult.

It was 1968; I was pregnant and married to a nightmare. We lived in Puerto Rico while my husband did research on brain functions. Two weeks after moving from the Bronx, my husband, Gil, turned into a raging lunatic. Periodically, he would slap or punch me; then, on bended knees, he would cry and apologize. I learned that if I didn't accept his apology, another beating would follow.

Stranded on a beautiful island, no phone, no car, except when he relented and let me use it, and pregnant, with complications, was as close to hell as I ever want to get. Whatever mistakes and screw ups I've made, that marriage and subsequent problems paid for any karma I've accrued.

It was after one of his insane apologies that we went looking for a dog. When I walked into the breeders home, a three month old puppy was brought out. Love at first sight is a habit with me, especially with animals. I had been reading Ayn Rand's *Atlas Shrugged* and Galt was the hero who suffered with a dignity I had never seen. He stirred my soul and I embraced the concepts of "riding the pain." So, Galt took a ride to Old San Juan. Mary and her little lamb never had it so good. He didn't need training. We were inseparable from the first. My husband hated the dog's being in our bed so as soon as he shut the door behind him to go to work, Galt jumped in and put his head on the empty pillow next to mine. Now Galt and Bunny were joined in my heart. Bunny seemed to have the same depth of devotion and sweetness. I ached for her. I ached for the lack of a Galt in my life.

Suddenly, aware of the voice on the other end, I answered, "The first thing I see is Bunny in Virginia. I see her being taken in a station wagon. There are two men in it. They feel like they just pulled a robbery. She is with a man with a shotgun. He is scuzzy looking. I see him on top of a hill."

"That's impossible. My car was stolen, but she's definitely not in Virginia. I live in Maryland, but I'm not near Virginia. That's quite a run. She can't be there. Please, I need your help. Can't you tell me anything else."

" You had a station wagon?"

"Yes."

" Then I do see it. It's the first town over the border, going directly south from you. I can only tell you what I see. Bunny is a large German Shepherd isn't she?"

"Yes, that's true."

"Then it's her I'm seeing. Call me back when you get any lead or just to talk. I'll see if more comes."

The call came the next day. Breathless with anxiety Sue began, "Bunny was seen in a station wagon heading south after a robbery with two men driving. The police think they were heading toward Virginia. How did you know? When will they find her? Is she okay?"

"Yes, Bunny is not hurt, but she'll be hungry and scared. I don't see the men harming her."

"How can you be sure?"

"I'm traveling there with my mind; that's how I can tell. Sue, I've been like this ever since I can remember. I'm so glad Bunny has you. You must love each other very much. Thanks for calling and keeping me informed. I'll be praying for her. Keep sending her loving thoughts. Keep your fears away from your thoughts of her. It's important. She's very telepathic; loving you, she is very sensitive to your thoughts."

"But how am I ever going to get her back? Where is she? She can't be in Virginia."

"Why don't you send me a photo of her. In the meantime, until I get it, please work on letting some of the fear abate and remaining open to the possibilities I'm seeing. I could be wrong, I know, but I could also be right. Call me in a day or two."

The photos were on my desk two days later when her call came.

"Hi, Sue. Okay, this time I see her by railroad tracks. Wait a minute; I think I feel a name coming...Harper's Ferry, West Virginia. Isn't that a movie or something?"

"Are you sure?"

"Yes, hold on there are more images coming. I'm waiting for them to clear."

The images were barely there. They were like looking through a long tunnel to see a speck of something, only you know the speck is the key to a mystery and it's important to get a detail shot and somehow enlarge it, translate it into common sense and be right.

"There are mountains and water nearby. She's on a mountain, alone. She's waiting for you. You can find her. Sue, you are going to see a truck, wooden sides, color green, and there's a bandanna on the seat."

"Is that where Bunny is?"

"No, but it's a clue to tell you that you are on the right track. I think she is by the railroad tracks near the mountain. Poor girl, she misses you. You've been sending her a lot of good messages. I can feel her being comforted."

Sue called back with the good news. Bunny was in her arms again.

I turned to Galt, now many years a spirit: "Thanks for keeping Bunny company; I love you. Just once more I'd love to hug you. You kept me company in some of my darkest moments."

Exercise: Believe the Instant Flash

Take index cards, a notebook, or a tape/message recorder and keep it with you at all times. The instant flash of intuition occurs in a seemingly random way. Just keep programming to be available. When you do notice a sudden out-of-the-blue image, thought, sensation, word, etc., note it down and record the date. Trust that it means something. You may have to interpret its meaning or it may be clear and direct. Even when it appears clear, it can still be a metaphor. Seeing death isn't always a literal clear vision; it can and usually does mean the end of something as you know it. Had I seen Innish, Pat's dog, while he lay near death, the vision of death would have to be interpreted to mean he is letting go of the life he had (or lack of) and moving into a completely new life, even in his physicality.

Christina is blind; Regal, her seeing eye dog, was lying at her feet when Christina sat in my animal communication class.

"How am I supposed to work with a photograph?" Christina wanted to know.

"No one gets to see what is in the photo; the rest of your classmates have their photos in envelopes so they cannot see."

The photo I handed Christina was of Beauty, the Morning Dove we had rescued when at about two days of age she had fallen out of a tree. In the photo, she was about 3 months old, standing on a cut off tree top on our lawn. It was her first solo flight.

Concentrating on quieting her chatter and just allowing herself to feel the energy emanating from the photo, Christina did as I asked.

When it came her turn, like many others she remarked, "I don't know, I couldn't possibly really know anything. I thought some things and here is the thought that popped up. First, it was of a bird. I don't know what kind, an uncommon rescue and a release."

Could she be anymore accurate? I doubt it. Many people discover when you remove the distractions and you can't use the excuse "body language" what's left is that invisible metaphysical universe of mind, thought and spirit.

Your turn. Start recording your instant flashes, learn to trust them.

Continue to record your instant flashes

More possibilities:

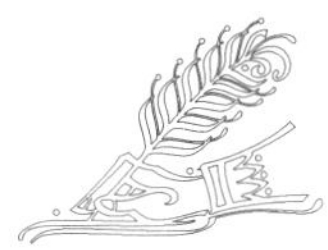

Look around your dwelling and find letters, note cards and objects others have given you. Sit down with a notebook or tape recorder at your side. Take each item and hold it while you enter a quiet, meditative state. Observe your thoughts, feelings, ideas. If words pop up, notice them, if images quickly flit by, note them too. Date and record all your findings. When you are ready check out your findings.

When anyone is missing,
They are missing from our view;

Find a way to enter into a
peaceful moment;

Create a view of your loved one,
Peaceful yet missing you;

Place this view in your heart;
Send this view to the Light;

Leave all outcomes
to our Creator.

Where's The Risk?

Thursday night, sometime in 1976. It was almost a year after I began teaching my first meditation, psychic/spiritual development class. A dozen or so regular clients had suggested that I gather them together on a regular basis. Another prediction had come true.

It was on one of those Thursdays' that Margaret, student, client and friend rushed into class late and breathless. Margaret has a face shaped by kindness and loss. A sweet sad look usually sat in her eyes, but tonight her eyes held troubled concern.

"Sorry I'm late, but one of my horses has a problem and I didn't want to leave the kids with it. I left Jane, the groom, there anyhow to see if she could calm him down."

A vision popped up in my third eye.

"Margaret, I see something. Let me show you on paper. I don't know the anatomy of a horse well enough to name the parts I'm seeing."

Grabbing paper and pen, I drew a rough outline of a horse. On the lower part of the back and to the right I put an X.

"This area is injured."

"I can call over to the barn and check on it. Can I use your phone? Jane, go over to Fellow and touch him and tell me what happens. Success: "Now what do we do? She touched him where you said and he jumped."

"Have her get some olive oil, warm it slightly and gently rub it over that area. Then, take the ends of the muscles and firmly pull them in opposite directions. Lastly, tell her to keep that area warm and when she is finished to place her palms down on the spot and imagine pouring light down from the universe, through her body, through her palms and into Fellow. While she is doing that, she can mentally send him a message of love and help, telling him to use the energy to help alleviate his suffering."

All made up? I had no idea what I was talking about; just kept following an image I was seeing and words I was hearing. Imagination is the act of imaging.

What is made up? Where does it come from? Who really knows? Not me, it's all made up. I figure I'll know for sure sometime after I leave my body permanently and then it may not matter.

The next week Margaret came in on time.

"Thank you, Nancy. Fellow is back to his old, wonderful self. Jane did exactly as you said and by the time she was finished, she could feel him relax as if he knew she was helping. Saved us a vet bill, too. Anytime you want a ride, just let me know."

Horses and I relate well as long as I am sending them healing and love. I always think they know I wish for them an existence where they are treated as if they have a soul, not as a commodity. Some of the folks I have met know the difference and have a lot of respect for their animal friends. Others treat them as if they are chattel, just as women, children, Asians, blacks, Jews and anyone standing out have been bought and sold at various times in our evolutionary process.

Exercise: Promote Your Feelings

When I was in Brooklyn College School Of Nursing, Mrs. Norman, my first teacher of nursing and the wisdom of caring, stated, "When entering the room of a sick patient, put your troubles in a bag at the door and leave them there. Pick them up when you leave if you still want to." It's been my exercise ever since, and ever since is many decades for me. If you haven't practiced coming from the Divine Center and setting the self aside, this is a great time to begin. We need you. If you are well on your way, thank you. You help make the world a more peaceful planet.

Practice creating a vision of you in situations that normally throw you. See yourself throwing all your fears, anger and pain in a bag and leaving it at your side. Now, continue your vision to a conclusion that is fear free, filled with your powerful wisdom and love.

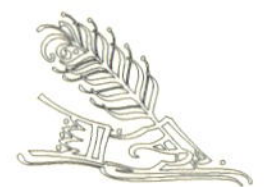

We can choose to enter
a world filled with
magic and miracles,
All we need do
is listen with our heart.

**Communication can occur
in a moment;
Awareness of
the communication
can take years.
Begin by making friends
with the unknown.**

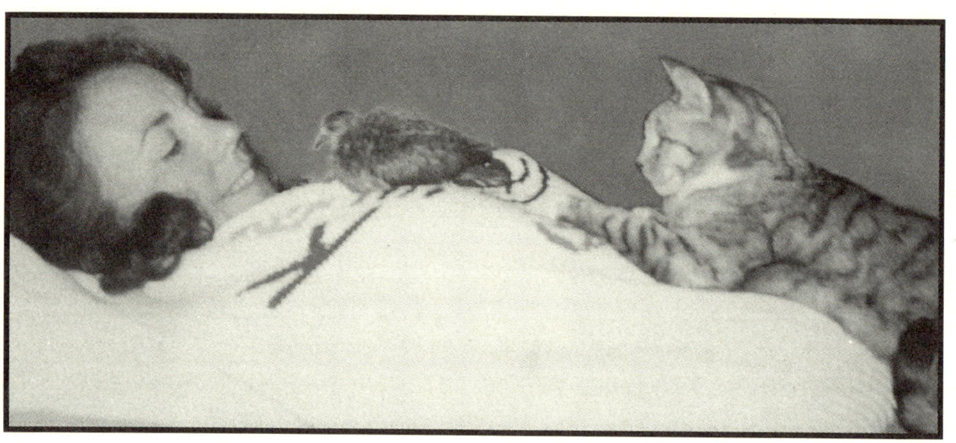

Nancy, Beauty and Tina

DONALD'S PUPPY

Late October, 1988

Sandy had a Doberman Pinscher who was 12 weeks old. She had him because of a friend. Sandy brought the little Doberman back from a Doberman show in California because she knew that her friend, Donald was going to die. He had cancer and it was just a matter of time.

Don was Sandy's closest friend and she knew that this was his last time at a dog show. He had seen this little puppy and said, "I just have to have that puppy." They were both professional dog handlers. Sandy hoped that by buying the puppy she would give Don the belief that he would be around in two years, which would be when they would be able to start to show the dog.

Sandy brought the dog home. She lives in the middle of nowhere, isolated, on nearly one thousand acres surrounded by the Pastiche river. The river suddenly stops its course and goes around, surrounding this acreage, and she lives at the head of the acreage. She had an older female Dobie who never got lost. So on a Sunday morning at 9:00 AM Sandy, her dog and the puppy took to the main roads so this little dog would know where he was. Sandy's faithful companion, Astin, who was ten years old, was running ahead with the little puppy following her.

Astin took them on the main trail. Sandy and Astin knew the land very well. When Astin made a right turn; Sandy knew where she would go and made the right turn off into the field after her. Sandy then saw her make another turn, with the puppy following, but they were quicker, and that was the last Sandy saw of them. She called and called. Finally, after an hour, she saw that it was useless and went back to her house. Sandy's husband, Joe said, "Astin's here." Sandy said, "I'm going to kill her, but at least they're back." Then Joe replied, "They're not back. Astin is back." By this time, it was about 11:00 AM. Sandy went out to look for the puppy again.

Three hours passed and she still had not found the puppy. It was winter, it was very cold and this was a very little puppy. He was not used to the cold or being outside. He had no idea where he was, and it was now 3:00 PM. Sandy panicked. She took Astin, and also rounded up quite a few other people with herding dogs and retrievers to help her look. Still no puppy. It was now night time.

Sandy was up all night long, calling the dog that didn't even know his name. She barked, howled and pleaded. She thought he might be scared into barking. He probably was scared to death, but he did not answer her barks or howls.

The woods were endless. Sandy kept looking with other dogs. When one got tired she would go back and get a different one. She had experiences with dogs becoming feral after just a few hours of being lost. They become frightened and disoriented, they don't even know you. She thought he would relate better to another dog, so she just kept searching with dog after dog. She had visions of the puppy starving and becoming weaker. He had plenty of water because it had been raining, at least.

Sandy called me on a Tuesday night. She had exhausted every other possibility. Then she remembered that a friend of hers knew someone with psychic abilities who had found a lost dog in New York City after the dog had been gone for about six weeks. She called her friend and was given my name and phone number. Sandy was hysterical. I told her that I saw a garage and willow trees. She said, "Willow trees don't grow wild, and I've been all through the woods, hundreds and hundreds of acres. There are only two willow trees and I planted them myself by my barn, which is 500 feet from the house." I continued telling her that I also saw sewers, pipes and ditches. She answered, "There ain't no such thing." I went on to tell her that I also saw a golf course. "Wrong," she said, "You're completely off base." Then I told her to let me think about this overnight and that I would get back to her in the morning. Sandy felt as though her last hope had vanished into thin air.

After we hung up, Sandy started thinking. There were no sewers, but the power company had come into these woods about ten years ago; and they had wanted to put up power lines. There was a river and they couldn't put up the lines. They came also to build a road, and what did they do? They brought great, big sewer pipes and threw them in what Sandy thought was a tributary of a river. They thought they could cover them with dirt and drive their trucks over them. Then Sandy said to herself, "Sewer pipes." She also realized that I had said ditches. Sandy went off to question one of her neighbors. Sandy's neighbor, who lived in an old farmhouse (this land was part of a farmstead about 100 years ago) said, "Let me explain it to you. When this was a farm, lots of times it went under water, so they had to dig ditches. They dug at least twenty of them. These ditches were dug from the high part to the low part. Then, they built a big ditch that went into the river. They are sewer pipes now." What Sandy thought was a river was really a drainage ditch. It was the middle of the night; it was pouring, and out she went to search around the ditches about a half mile from her house. Altogether she found four sewage drainage pipes.

Sandy was afraid the dog had drowned, so down she went, crawling through the ditch. Remember that it was just water, not a sewer. No dog. She went home, but as soon as the sun came up she went out again. She started following some side tributaries and followed them up toward the house, where they disappeared. They had been covered up for over 100 years, but she found them all. Sandy also looked for a golf course, found none, then thought, "Wait a minute. The main ditch goes out to the Pastiche River. Now, where this ends and all these little drainage ditches come down to it, across the river is a defunct golf course. They built a golf course and it went under water. There is a fairway right across the river." Sandy ran home to call me and tell me what she had found. I told Sandy, "The dog isn't far away. You are confusing him. Remember dogs don't think in English. They understand intonations. They can picture. If you believe in all the electrical impulses that bring birds home and dogs back, what you are doing is confusing to the dog. You're hysterical. There is this little thing out there, frightened, and if you are mentally sending anything, you are sending terribly frightening things. Clear your mind, concentrate on bringing him home. The things I keep seeing are the barn, willow trees and garage. Bring the dog back."

Sandy sat for hours, and when she went out she would also visualize the little dog and bring him to these trails. Once he found the main trail which led to her house, he couldn't get lost. It was Wednesday. Sandy called me. I told her, "I don't see death. He's very quiet, he's very near."

Nearly a week had passed; it was hunting season. Sandy asked any hunters she saw to please watch out for the dog and to try to bring him home. She called me again and I asked her to come to my house and bring me something the dog had slept on. Joe drove her to my house. By this time, she was nearly suicidal. She had no hope, she couldn't tell Donald. She was really at the end. She brought some little dog toys. I looked at her and said, "You have to go on with your life." I looked at Joe and said to him, "If you don't get her some help, I will. She's in very bad shape." This was a nightmare for Sandy. I told Sandy the dog would show up on Sunday. I told her to stop looking for him; he would just show up.

Joe promised he would get her to a psychiatrist and they left. When they arrived home, Sandy called the dog breeder whom they had bought the puppy from and told her the entire story. The breeder said she had another dog. "We won't tell Donald," Sandy and the breeder agreed. The dogs were identical. Maybe they would wait six months and then tell him, or maybe never tell him. She questioned how the breeder could sell her another dog when she had just lost one. The breeder said it was an accident and she knew she didn't do it on purpose. She said, "I'd like to tell you a story about accidents. You've known me about 18 years, but 23 years ago I had a little boy. My husband and I were going to a dinner and I let my sister and brother-in-law take my little boy for the night. He was two years old. In the morning my brother-in-law went for a walk with his German Shepherd and my son. They lived near a park with a lake. They were walking in the park on a cold winter morning. He heard a sound and turned around and the child was gone. He had fallen into the lake, under the ice. He died. Don't tell me about accidents. It wasn't anyone's fault, yet they are still blaming themselves for it." Then the breeder let

Sandy have the dog. Now it was Sunday, Joe was afraid to leave her alone, she was so distraught.

At 8:30AM Joe asked her if she would mind staying alone for a few minutes so he could go out to get a newspaper. She said she was not fine, but pulling herself together a bit more, so she thought it would be OK. He came back about 15 minutes later. He's a very laid back person; he doesn't get excited. He's quiet, gentle and easygoing. He walked into the house, voice raised and said to Sandy, "What the hell did you do?" Sandy did not understand and asked what the matter was. He replied, "I don't understand you. I don't believe you're this stupid. You lose a dog, we have spent the last seven days in total torture. Now, we get the other dog. We won't talk about what you had to pay for him; so, why would you let the new dog outside alone?" Sandy said, "I didn't let him out." Joe kept yelling, "Why did you let the new puppy out?" Sandy asked, "Out where? Where is he?" Joe answered, "Out in the garage. I drove into the garage and almost hit him. He was sitting there." Sandy was confused but knew there was no way the puppy could have gotten out. They went to the garage and the puppy was still sitting there. Sandy said, "Where is his ear tape? I just did his ears. My God, he's thin." She stared at him and said, "Wait a minute. I think the new puppy is still in the house, asleep. Let me check." Sandy ran into the house and yelled, "Puppy's in here." Joe then asked, "Well then what is this sitting in here? What do you mean he's in there? He's in here. I'm looking right at him." Sandy ran back into the garage with the second puppy and said, "Here he is." Joe asked, "Then what's this? This is weird, it's exactly seven days to the hour that the first puppy was lost. Now we have two puppies, one very thin!"

Sandy ran into the house to call me. Then she called Donald. Two weeks later she drove both puppies to Donald's house in Maryland so he could choose one. They were identical except the lost puppy had a totally different personality from the week before he was lost. He had become assertive, even with the other puppy. The other puppy had been the dominant one in the pack. Donald took the first puppy for a year and showed him. The dog was doing really well. Donald died and his widow was left with the dog. Sandy spoke to her every day. One day she told Sandy that something was wrong with the dog and she didn't know what it was. The dog was getting stranger every day. This was an 80lb, full grown Doberman.

Sandy went to Maryland to pick up the dog because she felt that he was having separation anxiety. He would rip up the room every time Donald's widow left.

Sandy and Joe now had the dog and his brother, but she had to keep them separated. He was definitely strange. When he was upset, he would whirl and whirl. Astin, the infamous surrogate mother had died. If the dog needed to go out and it was raining, he would howl and cry and shiver.

A year or so later I was the featured speaker at a kennel club. I spoke to a room full of people about my animal stories. I recognized Sandy and Joe in the audience. After my lecture Sandy stood up and asked if I ever get to meet the people or animals that I have found. I answered, "Very rarely." Sandy said, "Well, I have a treat for you."

Joe brought the Doberman out, a big, friendly dog. He turned him loose. He walked right to me, though we had never met, and stopped dead in his tracks. He looked at me; I patted him and didn't say anything. He walked away to visit some other people. I sat on the floor; suddenly the dog stopped as if something had happened. He whirled around, all four feet at once, then ran back over to me, climbed into my lap and put his head on my shoulder. Now, in dog language, this is totally submissive behavior. It was a body language he had never done with anyone before, Sandy told me. I don't know exactly how the dog knew me, but he knew me. It was like a complete submission to a mother. Sandy has told me in later conversations he has never again done this with anyone else.

Exercise: Believe Your Glimpses

You know what peripheral vision is, right? That quick, fleeting glimpse off to the side. In martial arts there is a term called *soft eyes*. It means do not look hard at something; you won't see it! Relax and look down with a soft feeling of looking and you will actually see more of the periphery. That's what we do with our third eye too. Look at the thought of a lost puppy or a sick pet. Look at it as relaxed as you possibly can. I know it isn't easy, but looking at it with fear and worry creates mixed results that you won't be able to rely upon.

The first thing I tell people who are frantically looking for a missing pet is to breath deeply and imagine you, your family, your home, your surrounding area and all else are encompassed by a wonderful, invisible, Divine Light that causes us all to be connected. Now, see it as having the divine intelligence and wisdom of the Creator of All Life. The Light will be able to travel, with your love, to your pet. You do not need to know how or where. You need to believe and have faith in the process. Then, sit a few more minutes and keep breathing deeply. Hold in your mind and your heart the sweetest moments between you and the missing animal. Send these sweet memories along the Divine Light. *Now watch for any thoughts, images and ideas.* You need to be in universal state of mind when you glean messages. They are glimpses into the Light of All Life. Use the glimpse however odd or kooky it seems. It may seem that way because you are not the animal that is missing. Coming from our rational train of thinking doesn't often help. Here is a letter from a client that might help you.

"On March 28, 1999 my 4 year old toy poodle Taco ran away. He was a very nervous, scared dog who never left my side, except to sit on the lap of one of the five people he trusted in the world. It was unusual for him to take off on his own.

My husband and I were going through a divorce. I think Taco was as anxious and confused as we were. During this time, we still shared a love for Taco. So, together, with the help of friends and family, we started searching for him from sun up to sun down. At night, I would cry knowing Taco was alone, wondering if he had shelter and food, hoping my love for him would protect him and keep him safe.

By day two all our efforts had failed and a good friend suggested I call Nancy Weber. She would have the wisdom, guidance and insight that we needed to find Taco.

The Rangers at a park three miles from where Taco lived saw him once a day. When they approached him, he ran. While talking to Nancy, she assured me that when domestic animals take off on an adventure they have the ability to adjust to the wilderness. Taco would be capable of finding scraps of food and shelter. She said a lost dog typically moves in a large circle, spiraling in towards home or wherever they hope is home. He would continue to roam around this circle for several days. I should place a familiar toy, piece of clothing, etc. along the path where he was spotted. I felt comforted after talking to Nancy and thought there was a real chance of finding Taco.

Knowing all this, we decided to leave a truck that Taco spent a lot of time in at the park, overnight, with the door open. On the morning of day five, Taco was found sleeping on the seat of the old truck. This was one of the happiest mornings of my life.

Two years passed and I was running with my girlfriend when a woman in her car stopped to ask if we've seen her lost dog, Emily, a golden retriever. I tried to help the woman with the information I had learned from Nancy. She was too upset to listen. My friend and I decide we would extend our experience of finding Taco to trying to find Emily. We could be more logical in our effort without the anxiety of the lost dog belonging to us. Within three days, we found Emily, persuaded her to come to us with a piece of cheese and collared her for the people who missed her, loved her and needed her in their life.

And the beat goes on,
Love, Robin

Focus on being an instrument of
Divine Light.
Trust the very first glimpse,
idea or feeling.
Record it.
Keep doing this.
Now, help the world.

"Out for a stroll"

SHEBA'S LOST

Iris Nevins' has a pack of five Shelties who live a life of canine luxury on a New Jersey farm. They have three different fenced in yards, totaling about five acres, where they can run and have adventures to their hearts' content. They also have a small old barn for shelter, which is divided into two rooms; it may well be the largest "dog house" in the whole state. When the sun goes down, the dogs come into the main house and spend the evening with the family, sharing tidbits of dinner and fall asleep wherever they like. The three males are very protective and sleep in the bedroom at night to keep watch over Iris.

One day, Leslie, a friend of Iris', was dog sitting for her friend Janet's Sheltie, "Sheba". Tuesday, August 23, 1994, Iris received a call from Leslie. She had to go out of town and needed someone to watch Sheba and her own dog, Tanner. Iris and Tanner are best friends. Whenever she would visit, Tanner would hear her car and come running up howling, "I wuw woo, I wuw woo!" Then Iris would calm him down, and tell him that she loved him, to which he would reply, "I wuw woo woo". Unfortunately Iris' three males did not like other males visiting, so Tanner stayed home. Iris took Sheba home, an adorable little female.

Iris had met Sheba a few times before, but she was a nervous dog, and when Leslie brought her over she was a bit frightened by Iris' howling pack of dogs. They were all very curious and it was apparent that Sheba was not enjoying having five other Shelties sniffing at her. Leslie drove off and Iris took Sheba inside the house to give her some leftover meat. Sheba followed her inside.

Later that day, Iris put Sheba and her two females in a separate yard and they seemed to get along peacefully. She watched the three female dogs for the next hour, bringing more meat out to them several times, making sure Sheba was OK. Then, Iris had to go out for the afternoon. As she drove her car down the driveway, Sheba ran after her along the fence. Iris got out at the end of the drive and reassured her that she would be back in a few hours.

When Iris came back later that afternoon, Sheba was gone. Iris and some friends searched all over the huge yard, behind trees and bushes and only found one little spot under the fence where she could have possibly tunneled under. She may also have climbed the fence. We will never know for sure how she got out.

Iris called every police station, veterinarian, shelter and animal control person in our own county and every one surrounding. Sheba's owner was off hiking in the woods in Maine and would not be available by phone so Iris left a message on her answering machine in Pennsylvania, hoping she might call home for her messages. She even called three radio stations in the area and had them make announcements several times a day.

Then Iris called me. Immediate impressions of a Sheltie digging under a fence to find her way home landscaped my mind. "I see that someone will pick her up in two days and bring her back to your house. Right now, I see her at an abandoned railroad bed."

I've known Iris as a dear friend for a long time and knew that she wouldn't stop the search until Sheba was found. I reiterated that she would not find Sheba, but her search would help the person who finds her.

Iris hung up and went to the only abandoned railroad bed she knew of in the area and started to holler "Sheba". It was Tuesday night. She took Buster, the smartest male in the Sheltie pack, out looking for Sheba. If Sheba was around, Buster would find her. It was getting dark. There was no sign of Sheba anywhere. She stopped people on bicycles and asked them, but no one had seen her. She was a shy dog and Iris was afraid she would stay well hidden from anyone who was not familiar.

The next two days, Iris continued her search. It was not until late Wednesday night that she heard from Janet, Sheba's owner. She was very understanding and told Iris not to blame herself. She said she would drive from Maine the next morning and help look for Sheba. Iris called me a few more times, and I told her I felt Sheba was pretty close by.

When Janet arrived Thursday afternoon at about three o'clock, Iris was a mess. She had hardly eaten or slept since Tuesday. It's bad enough when you lose your own dog, but it's so much worse when it belongs to someone else. Janet called all through the fields and woods, still no Sheba. Janet and Iris decided to get in the car and drive to Leslie's house to see if she made her way back there. As they drove, Iris told Janet about my comments and that I had told Iris to mentally tell Sheba to go to a place where people would see her so she could be picked up and brought back.

A few hours later, they were still driving around. Iris' car phone rang. Iris' daughter shouted elatedly, "A trooper just brought Sheba home!" Iris asked her if she was sure it was Sheba and not another sheltie, and she said she was sure.

As it turned out, a trooper from the local police barracks, the first person Iris had spoken to, was the one to find her. She was picked up about three miles from Iris' house in a small town. The trooper saw a dog that looked like a fox and remembered the description. She wouldn't come to him when he called her name; so, he parked his car, opened the back door and sat down in the driver's seat. A minute later, Sheba, who is very used to riding in cars, jumped right in. Just two days later and someone did pick the dog up and bring her back, someone that Iris had spoken to in her search.

If you are going to assist in finding missing animals, let me share a few other experiences with you. The first great lesson came in the mid 1980's with a phone call from a man whose dog "never left the yard." That is probably the most repetitive statement I hear; the shock at the sudden change of behavior. While we are friends with and may live with cats, dogs, birds, gerbils and others, let us all remember they are different species. A dog who never has wandered from home may, even after years of wonderful companionship, suddenly decide it is time for an adventure. It is not done to harm or hurt anyone. A dog is exercising its dog rights to be a dog!

Jack, a Terrier, was five years old when Doug called me, frantically worried about his beloved pet. He was also devastated that Jack would want to run away. As soon as he spilled his fear out, my mind shot an image of a map. I zoomed in, not always easy. Sometimes, if I try to change the image, it goes away. This time, the zoom worked and I saw Jack due north from home exactly one town away. Doug raced to the town and posted flyers containing all the necessary information including an excellent photo of Jack. He talked to dozens of people, left posters at the animal shelter, animal hospitals, police station, schools and any retail stores that were willing to post the flyer. Days later, nothing. Doug went back to the town and again spoke to people. He found several people on one street who all saw Jack but thought he was just a local dog out on a stroll. They had seen the poster but it never sank in. Doug was baffled. He called me again.

"Okay, now I see Jack further north."

Rather than tell you each step, just reread the above paragraph. Yes, people spotted Jack. Yes, they ignored the posters. Over a period of one year, Doug and I spoke every week. We began to notice that my vision and the confirmation of each town was showing an interesting picture of a spiral. Jack had, for some reason, gone north then east, south and west. Each move, he was spotted a little closer to the epicenter. Each town, the story was the same, Doug would rush there, post flyers and ask questions. Days later, he would return and find that while I was correct, no one would think to call. Almost a year went by when we both noticed that I saw Jack within a few miles of home. Exactly a year to the day after he left, Jack showed up on Doug's doorstep wagging his tail, eager to share his love. Jack appeared healthy and quite happy. He continued his life with Doug as if nothing had changed. Jack never ran away again. Doug finally accepted that Jack was not a child but an adult dog who had decided to see the world. Doug gained a lot of respect for Jack's ability to care for himself, enjoy his adventure and still be devoted to his best friend.

Exercise: Everything is useful

Throughout the year of searching for Jack, Doug was very gracious and respectful of my work. His attitude made it easy for me to continue and not be fearful of the outcome.

Even being wrong can be helpful, maybe not to those we work with at the moment, but it can stretch us into greater understanding. A woman lost her dog and brought me a photo of Maisey. All I could see was emptiness and death, as if the dog no longer existed. Knowing that a vision of death can be a metaphor, I carefully explained what I saw could be symbolic but I couldn't see any further. The woman called three days later to tell me the dog had been found alive but injured. Maisey had a seizure when she was found.

The vision was definitely about a major physical change. I didn't see the trauma and I didn't feel Maisey alive. What happened? I stopped at a level that wasn't deep enough. Feeling uncomfortable with the image of possible death, I let my fears take over. Fear has its own chemistry and will shut down the intuition and psychic input faster than you can blink. It's difficult to accept that while working on a situation, only hindsight helps put it together. Insight is staying open; hindsight indicates I've had a moment frozen with fear. Getting comfortable with discomforting ideas, such as mortality, is mandatory for clarity. I keep working on it, and instances where I'm wrong lead me to a greater understanding of how my fears operate.

When did you doubt your intuition? Have you forgiven yourself? Have you taken the lessons and applied them to other situations?

We all have to learn how to say good-bye. Carol had lovingly said good-bye to Natasha many months after this letter arrived.

Dear Nancy,

I want to thank you for the session that you did this past April 3 with my dog, Natasha, and myself. Before the session, I felt that the end of her life was near and was seeking help in saying good-bye to her. During the session you communicated to me for her that she was NOT ready to go, yet.

After the session, she really "came back to life" (ironically on the day before Easter!!) and we have had (and continue to have) a richer, deeper connection than ever before. And she wasn't the only one that changed!! She continues to suffer from the crippling Degenerative Myleopathy; she has good times and bad times and is progressively but slowly losing her physical abilities. I have learned so much about how to care for her physically AND nurture her spirit (and let her nurture mine). I greatly appreciate what you did for us.

Thank you,
Carol

May your own journey with this beloved planet of ours be filled with hopes, memories and dreams.

May you have the great joy of spending many moments of your life with friends of all walks - literally and figuratively.

And may you find the incredible richness of an existence filled with communing with all of life.

With Light filled Love,

Nancy Orlen Weber

Dick Weber and his hairstylist, Samantha

Dick Weber and Samantha cat napping

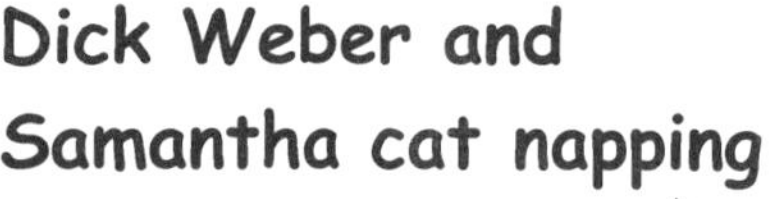

The author's buddies
Kit, Kaboodle and Samantha

To order your copy of "PSYCHIC DETECTIVE"
1-888-266-9462

Conscious Living Journal ~ Summer 2002

MIND TREATS

BY SONYA OPPENHEIMER

BOOK REVIEW

ISBN 0-9646118-2-1

Psychic Detective
Stories & Exercises for the Soul
by Nancy Orlen Weber

Unlimited Mind Publication

Who Done it? Have a hunch? Not because a skilled writer has cleverly crafted the story to allow you the delight of solving the mystery before the story's conclusion. Your hunch deals with real life, with breaking news, with events not being contrived for entertainment. And, in your mind's eye, you see. Are you just being fanciful or do you have valid insights, intuitive knowledge?

Because "all life is a creation…an exploration that has infinite possibilities," author Nancy Weber not only tells us that the latter is the case, she asserts that with simple exercises, each day "teaching your mind what you expect it to do," we can develop our latent psychic abilities.

In Nancy Weber's newly released book, "Psychic Detective Stories & Exercises for the Soul," she relates her amazing, true tales of psychic detective work and then, through exercises and analogies, helps the reader develop similar skills.

To start, Nancy writes about the simple delight of helping people find lost articles. Imagine the pleasure – and smug self-congratulations – of being able to tell your spouse where to find the missing car keys. Sure, all the logical places have already been searched. That's why it's your intuition that must be exercised into astounding success. It will be in the very first chapter.

Of course, Nancy Weber wasn't awarded a detective's badge by a New Jersey police department for finding lost keys. Tales of her work include heart-warming stories of finding missing children and blood-curdling ones of being called in to aid in unsolved murders. We feel her self-restraint as many negate the possibility of psychic powers and celebrate as she demonstrates her exceptional talent. And, then, wonder about our own hidden potential as Nancy outlines how she did it and provides step-by-step building blocks for developing our own individual powers.

Ironically, Nancy points out that even though most detectives recoil at the thought of anything other than cold, hard, scientific logic being the path to solving a crime, the best detectives are those who follow their hunches. Most would flinch at the notion of that being psychic ability but, as Nancy points out, call it what you want, anything is just another name for psychic ability.

"Psychic Detective" offers unusual multi reading pleasures. First, this book is a compelling read for anyone who enjoys a story well told. Then, mystery genre and true-story weave their irresistible pull on the imagination. All of this is sufficient enjoyment potential to highly recommend this book. But, like a fabulous desert after a wonderful meal, "Psychic Detective" also is a fun filled adventure in developing one's own intuitive powers. Just imagine the amusement when you are able to help family and friends find missing items.

About the Author:

Nancy Weber is a psychic, a registered nurse, an educator, a motivational speaker, a holistic health lecturer, a radio and TV host, an interfaith minister, an assistant to detectives and an author of prose, poetry and song lyrics. Co-founder and president of the Holistic Alliance International, she practices what she preaches and leads a life of fulfilling her unlimited possibilities-a potential with which she believes we are all gifted.